Environmental Exposure to Microplastics: Effects on Animals and Human Health

Environmental Exposure to Microplastics: Effects on Animals and Human Health

Guest Editor

Lei Su

MDPI

Basel • Beijing • Wuhan • Barcelona • Belgrade • Novi Sad • Cluj • Manchester

Guest Editor
Lei Su
College of Oceanography and
Ecological Science
Shanghai Ocean University
Shanghai
China

Editorial Office
MDPI AG
Grosspeteranlage 5
4052 Basel, Switzerland

This is a reprint of the Special Issue, published open access by the journal *Toxics* (ISSN 2305-6304), freely accessible at: https://www.mdpi.com/journal/toxics/special_issues/F1K648YC8D.

For citation purposes, cite each article independently as indicated on the article page online and as indicated below:

Lastname, A.A.; Lastname, B.B. Article Title. *Journal Name* **Year**, *Volume Number*, Page Range.

ISBN 978-3-7258-6242-9 (Hbk)
ISBN 978-3-7258-6243-6 (PDF)
https://doi.org/10.3390/books978-3-7258-6243-6

Contents

About the Editor

Lei Su

Lei Su is a researcher at the College of Oceanography and Ecological Science, Shanghai Ocean University, where he focuses on marine plastic pollution and its environmental impacts. His work primarily addresses microplastic and marine debris contamination, with research interests covering the transport processes of plastic debris and micro–nano plastics across different environmental media, their biological effects on marine organisms, and natural fragmentation mechanisms. In applied research, he pays particular attention to plastic pollution in aquaculture environments and associated ecological risks, while exploring the potential of artificial intelligence methods for pollution prediction and risk assessment. His work typically combines field observations with laboratory experiments to better understand the pathways and impacts of plastic contaminants in marine ecosystems. The application of AI technologies in his recent studies represents an ongoing effort to develop more effective tools for environmental monitoring and risk analysis. Lei Su's research contributions have been acknowledged through his inclusion in the Clarivate Global Highly Cited Researchers list for five consecutive years since 2021, reflecting the recognition his work has received within the scientific community. He continues to work on understanding plastic pollution processes and their environmental implications, contributing to the scientific understanding of this challenging environmental issue through persistent research efforts and collaborative projects with fellow researchers.

Preface

It is with great pleasure that I introduce this Reprint, "Environmental Exposure to Microplastics: Effects on Animals and Human Health". The burgeoning field of microplastic research has evolved dramatically from its initial focus on environmental occurrence to a more profound and complex interrogation of its biological consequences. This volume is born from a pressing need to synthesize contemporary scientific inquiries that bridge the gap between mere environmental detection and a mechanistic understanding of the impacts on biological systems, culminating in a perspective that embraces the interconnected One Health paradigm. My primary motivation in curating this Special Issue was to assemble a cohesive body of work that moves beyond cataloguing pollution and directly addresses the cause-and-effect relationships, exposure pathways, and potential health risks associated with microplastics and their associated contaminants.

The scope of this Reprint is intentionally broad, reflecting the multifaceted nature of microplastic pollution itself. It encompasses a spectrum of investigations, ranging from controlled laboratory studies elucidating toxicity at molecular and organismal levels to field-based observations that document real-world exposure and ecological distribution. The research contained herein explores a variety of model organisms—including zebrafish, rats, and avian species—to provide a comparative view of biological impacts. A central theme that unites these studies is the critical examination of combined exposure scenarios, acknowledging that microplastics in the environment rarely exist in isolation but instead interact with a cocktail of chemicals, from plasticizers like phthalates to pharmaceuticals and personal care products.

A key aim of this collection is to highlight the "carrier effect" of microplastics, a phenomenon where their surfaces facilitate the transport and enhance the bioavailability of co-pollutants, thereby amplifying toxicity. Furthermore, several contributions rigorously demonstrate that the intrinsic properties of microplastics, most notably their polymer type, are fundamental determinants of their biological fate and effects. This challenges the notion of microplastics as a single, uniform contaminant and calls for a more nuanced approach in risk assessment. Alongside toxicological mechanisms, this Reprint also showcases innovative biomonitoring strategies, such as the non-invasive analysis of bird pellets, which offer ethical and efficient tools for tracking environmental pollution and its impact on wildlife.

This Reprint is addressed to a wide audience of scientists, graduate students, policy-makers, and environmental professionals who seek a deeper, evidence-based understanding of the microplastic challenge. I hope it will serve not only as a valuable summary of the current state of the science but also as a catalyst for future research. The findings compellingly argue for the integration of aging processes, long-term chronic exposure assessments, and a greater focus on nanoplastics into future investigative frameworks. By presenting this curated compilation, my ultimate purpose is to contribute meaningfully to the scientific foundation required for developing effective mitigation strategies and informed policies to safeguard ecosystem integrity, animal health, and, ultimately, public health.

I extend my deepest gratitude to the authors for their excellent contributions, the dedicated reviewers for their invaluable time and expertise, and the editorial team of the journal for their unwavering support in bringing this important collection to fruition.

Lei Su
Guest Editor

toxics

MDPI

Editorial

Multiple Effects, Pathways, and Potential Health Risks from Environmental Microplastic Exposure

Lei Su

College of Oceanography and Ecological Science, Shanghai Ocean University, Shanghai 201306, China; lsu@shou.edu.cn

1. Introduction

After nearly two decades of extensive research, microplastics (MPs) have been documented in virtually all ecosystems and across diverse environmental compartments [1,2]. The focus of MP research has expanded from its origins in marine pollution to the current One Health framework [3,4]. A growing body of evidence now indicates that exposure to MPs and their associated nanoplastics may pose potential risks to human health, such as increasing the incidence of cardiovascular diseases [5,6]. Consequently, determining the actual toxic effects and ecological risks of MPs represents one of the most critical research priorities. In real-world environments, MP exposure occurs through multiple pathways, and their toxic effects vary depending on their intrinsic properties—such as size, polymer type, and morphology—in addition to the exposed species [7,8]. Therefore, assessing the toxicity of MPs within authentic environmental contexts is fundamental to evaluating their potential ecological and human health risks.

In light of the limited understanding of the toxicity and mechanisms of action of MPs with different characteristics, particularly under real-world environmental conditions, we organized the following Special Issue, "Environmental Exposure to Microplastics: Effects on Animals and Human Health." This collection comprises nine articles, including eight original research papers and one review. The articles included in this Special Issue cover a wide range of topics, from toxicological exposure experiments to field observations in environmental media and biomonitoring. They comprehensively address the toxic effects and potential risks of MPs to various organisms, including fish, birds, and mammals. Furthermore, several contributions address scenarios closely linked to human activities—such as wastewater treatment plants and the use of plastic products—thereby extending the discussion on ecological risk to the domain of human health. Collectively, this Special Issue provides valuable evidence and scientific perspectives for understanding the ecological impacts, human health risks, and exposure pathways of MPs.

2. An Overview of Published Articles

This Special Issue brings together nine articles that provide critical insights into the multifaceted impacts of MPs. Collectively, the research demonstrates that MPs act as carriers for co-pollutants like phthalates and pharmaceuticals, significantly enhancing their toxicity in aquatic and mammalian models. The study findings further reveal that the toxicological effects are highly dependent on polymer type, with polystyrene (PS) often exhibiting greater adverse effects than polyvinyl chloride (PVC). Fieldwork from diverse geographic regions confirms the widespread ingestion of MPs by wildlife, while innovative biomonitoring methods using bird pellets offer a non-invasive tool for tracking environmental pollution. A key finding underscores the often-overlooked pathway of human exposure through the mechanical fragmentation of plastic products during daily use. Lastly, investigations into wastewater treatment plants show that while MP removal is efficient, residual endocrine disruptors pose a persistent ecological risk. In the following sections, we systematically elaborate on these main findings.

2.1. Toxicological Mechanisms and Combined Exposure Effects

A cornerstone of this issue lies in elucidating the toxicological effects of MPs under environmentally relevant scenarios, particularly through combined exposure studies. The findings of studies by Zhang et al. (Contribution 3) and Wang et al. (Contribution 4), employing zebrafish models, reveal a critical synergy between MPs and co-pollutants. They demonstrate that polyethylene terephthalate (PET) and polystyrene (PS) MPs can act as carriers for Di-butyl phthalate (DBP) and methamphetamine (METH), respectively, enhancing the bioavailability and thus exacerbating the developmental toxicity and behavioral impairments induced by these chemicals. This "carrier effect" is further nuanced by the work of Xu et al. (Contribution 9), who highlight that joint toxicity is highly polymer-dependent. In their study, the authors found that while METH synergistically enhanced the toxicity of PS MPs, it antagonized the effects of polyvinyl chloride (PVC) MPs, underscoring the importance of considering plastic polymer types in risk evaluation. Complementing these aquatic toxicology studies, Gad El-Karim et al. (Contribution 1) investigated the mammalian system, showing that the plasticizer DEHP induced significant hepato-nephrotoxicity in rats through oxidative stress and inflammatory pathways, which could be effectively mitigated by the natural carotenoid lutein, pointing to potential intervention strategies.

2.2. Environmental Distribution and Biomonitoring

Beyond the laboratory, several studies provide vital insights into the environmental occurrence and distribution of MPs, forming the basis for ecological risk assessment. Ji et al. (Contribution 2) conducted a comprehensive survey in the Yangtze River Estuary, documenting the abundance, composition, and seasonal variations of MPs. In their work, they identified terrestrial input and hydrodynamics as key drivers of MP distribution, ultimately concluding that there is currently low ecological risk in this particular region. Extending monitoring efforts to biota, Bilal et al. (Contribution 7) and Bjedov et al. (Contribution 6) employed birds as bioindicators. Bilal et al. quantified MPs in the digestive tracts of ducks from the Panjkora River in Pakistan, confirming widespread ingestion and highlighting fragments as the dominant shape. Similarly, Bjedov et al. pioneered the non-invasive analysis of white stork pellets, successfully identifying a diverse range of anthropogenic particles and linking their composition to local industrial and waste management practices, thereby validating avian species as powerful sentinels for environmental plastic pollution.

2.3. Human Exposure Pathways and Risk Assessment

The pathway from environmental contamination to human exposure is a critical focus of this issue. Tong et al. (Contribution 8) directly addressed this link by investigating the fate of MPs and endocrine-disrupting chemicals (EDCs) in two wastewater treatment plants (WWTPs) in Shanghai. While WWTPs showed high removal efficiency for MPs, the persistence of EDCs in the effluent posed a non-negligible risk to receiving aquatic ecosystems. In addition, in the review by Yu et al. (Contribution 5), the authors identified a significant but often overlooked human exposure route: the mechanical fragmentation of plastic products during daily use. They argued that the release of secondary micro- and nanoplastics from items such as food packaging and textiles constitutes an acute exposure pathway, the magnitude of which may far exceed that from environmental sources, calling for an integration of product life-cycle and aging processes into risk assessment frameworks.

2.4. Conceptual and Methodological Advancements

Lastly, this Special Issue highlights conceptual and methodological advancements. In the review by Yu et al. (Contribution 5), the authors not only identify hidden exposure pathways but also establish a forward-looking agenda for the field. Together with the novel biomonitoring techniques demonstrated in the bird pellet and duck studies, these works push the boundaries of how we track and evaluate MP pollution, emphasizing the need for non-invasive, ethical, and comprehensive monitoring strategies that reflect real-world exposure scenarios.

3. Future Research Perspectives

This collection has advanced our understanding of MP toxicity and ecological risks by integrating findings from diverse case studies. Nevertheless, the toxicity of environmentally relevant MPs, particularly nanoplastics, has yet to be fully determined. Addressing these critical knowledge gaps demands technological innovation and a strategic evolution in research focus. The following key areas must be prioritized in future work to accurately assess the ecological and human health risks of MPs:

3.1. Inclusion of Nanoplastics in Human Health-Related Environmental Monitoring

Although MP surveys have become routine in some regions, nanoplastics have not yet been incorporated as monitoring parameters in environmentally relevant matrices closely associated with human exposure, such as drinking water. Due to their minute size, nanoplastics are more likely to penetrate human tissues and organs through environmental exposure, thereby elevating potential health risks. Moreover, once nanoplastics enter the muscular tissues of organisms, they are more prone to genuine bioaccumulation and trophic transfer, eventually reaching humans through the food web. Once technical bottlenecks in nanoplastic detection are overcome, priority should be given to establishing pollution baselines in key human-related media—including drinking water, food, and air—to support subsequent standard setting and risk management.

3.2. Risk Assessment of Low-Concentration and Long-Term Exposure

In real-world environments, MP exposure is generally characterized by low concentrations and long duration. However, many laboratory studies involve the employment of exposure doses significantly higher than environmental levels, which may lead to overestimation of actual risks. In addition, given the widespread use of plastic products and insufficient recycling, organisms and humans are continuously exposed to plastics through multiple pathways over extended periods. In current toxicological and medical research, long-term observational studies should be established to investigate the chronic effects of MPs, including potential intergenerational risks. There is also a need to develop more sensitive biomonitoring tools to enhance the detection of MP-related effects and enable early warning of exposure risks.

3.3. Considering the Impact of Aging and Weathering on Plastic Toxicity

Plastics in real-world environments are consistently subject to aging and weathering, which significantly alter their physicochemical properties. Firstly, the aging process leads to plastic fragmentation, generating secondary microplastics and thereby increasing environmental exposure concentrations. Secondly, aging modifies the surface characteristics and structure of plastics, potentially promoting the development of surface biofilms, which may enhance the likelihood of ingestion by organisms. As the structure of MPs changes during aging, plastic additives may also be gradually released. Certain additives with known endocrine-disrupting effects can form combined exposures with MPs under natural conditions. In future studies, aging and weathering processes should be regarded as critical environmental factors affecting MP behavior and toxicity, representing an essential attribute of real-world exposure scenarios.

Data Availability Statement: The data presented in this manuscript will be made available by the authors upon request.

Acknowledgments: The Guest Editor would like to express their sincere gratitude to all of the authors for their valuable contributions to this Special Issue, in addition to the reviewers for their time and expertise in evaluating the manuscripts. The editorial team of *Toxics* is also acknowledged for the administrative and technical support they provided throughout the process.

Conflicts of Interest: The author declare no conflicts of interest.

List of Contributions:

1. El-Karim, D.; Lebda, M.A.; Alotaibi, B.S.; El-kott, A.F.; Ghamry, H.I.; Shukry, M. Lutein Modulates Oxidative Stress, Inflammatory and Apoptotic Biomarkers Related to Di-(2-Ethylhexyl) Phthalate (DEHP) Hepato-Nephrotoxicity in Male Rats: Role of Nuclear Factor Kappa B. *Toxics* **2023**, *11*, 742. https://doi.org/10.3390/toxics11090742.
2. Ji, X.; Yan, S.S.; He, Y.L.; He, H.S.; Liu, H.Q. Distribution Characteristics of Microplastics in Surface Seawater off the Yangtze River Estuary Section and Analysis of Ecological Risk Assessment. *Toxics* **2023**, *11*, 889. https://doi.org/10.3390/toxics11110889.
3. Zhang, Q.; Ma, W.J.; Zhu, J.M. Combined Toxicities of Di-Butyl Phthalate and Polyethylene Terephthalate to Zebrafish Embryos. *Toxics* **2023**, *11*, 469. https://doi.org/10.3390/toxics11050469.
4. Wang, H.; Xu, J.D.; Yuan, Y.; Wang, Z.L.; Zhang, W.J.; Li, J.A. The Exploration of Joint Toxicity and Associated Mechanisms of Primary Microplastics and Methamphetamine in Zebrafish Larvae. *Toxics* **2024**, *12*, 64. https://doi.org/10.3390/toxics12010064.
5. Yu, Y.; Craig, N.; Su, L. A Hidden Pathway for Human Exposure to Micro- and Nanoplastics-The Mechanical Fragmentation of Plastic Products during Daily Use. *Toxics* **2023**, *11*, 774. https://doi.org/10.3390/toxics11090774.
6. Bjedov, D.; Mikuska, A.; Gvozdic, V.; Glavas, P.; Gradecak, D.; Bogojevic, M.S. White Stork Pellets: Non-Invasive Solution to Monitor Anthropogenic Particle Pollution. *Toxics* **2024**, *12*, 236. https://doi.org/10.3390/toxics12040236.
7. Bilal, M.; Yaqub, A.; Hassan, H.U.; Akhtar, S.; Rafiq, N.; Shah, M.I.A.; Hussain, I.; Khan, M.S.; Nawaz, A.; Manoharadas, S.; et al. Microplastic Quantification in Aquatic Birds: Biomonitoring the Environmental Health of the Panjkora River Freshwater Ecosystem in Pakistan. *Toxics* **2023**, *11*, 972. https://doi.org/10.3390/toxics11120972.
8. Tong, Y.X.; Xie, M.J.; Xv, H.; Sun, R.H.; Wang, Q.; Li, J.Y. Microplastics and Endocrine Disruptors in Typical Wastewater Treatment Plants in Megacity Shanghai. *Toxics* **2024**, *12*, 345. https://doi.org/10.3390/toxics12050345.
9. Xu, J.D.; Yang, W.Q.; Wang, D.Y.; Wang, Z.L.; Liu, C.; Li, J.A. Methamphetamine Shows Different Joint Toxicity for Different Types of Microplastics on Zebrafish Larvae by Mediating Oxidative Stress. *Toxics* **2024**, *12*, 9. https://doi.org/10.3390/toxics12010009.

References

1. Thompson, R.C.; Olsen, Y.; Mitchell, R.P.; Davis, A.; Rowland, S.J.; John, A.W.; McGonigle, D.; Russell, A.E. Lost at sea: Where is all the plastic? *Science* **2004**, *304*, 838. [CrossRef] [PubMed]
2. Sheriff, S.S.; Yusuf, A.A.; Akiyode, O.O.; Hallie, E.F.; Odoma, S.; Yambasu, R.A.; Thompson-Williams, K.; Asumana, C.; Gono, S.Z.; Kamara, M.A. A comprehensive review on exposure to toxins and health risks from plastic waste: Challenges, mitigation measures, and policy interventions. *Waste Manag. Bull.* **2025**, *3*, 100204. [CrossRef]
3. Procopio, A.C.; Soggiu, A.; Urbani, A.; Roncada, P. Interactions between microplastics and microbiota in a One Health perspective. *One Health* **2025**, *20*, 101002. [CrossRef] [PubMed]
4. Wright, S.L.; Kelly, F.J. Plastic and human health: A micro issue? *Environ. Sci. Technol.* **2017**, *51*, 6634–6647. [CrossRef] [PubMed]

5. Marfella, R.; Prattichizzo, F.; Sardu, C.; Fulgenzi, G.; Graciotti, L.; Spadoni, T.; D'Onofrio, N.; Scisciola, L.; La Grotta, R.; Frige, C.; et al. Microplastics and Nanoplastics in Atheromas and Cardiovascular Events. *N. Engl. J. Med.* **2024**, *390*, 900–910. [CrossRef] [PubMed]
6. Liu, L.; Tu, P.; Niu, H.; Li, X.; Gong, X.; Chen, Z.; Xing, M.; Wu, L.; Lou, X. Cellular Impact of Micro(nano)plastics on Human Health: A Review. *Toxics* **2025**, *13*, 913. [CrossRef]
7. Kim, L.; Kim, H.; Song, Y.; An, Y.-J. Chronic effects of irregular and fibril microplastics on *Artemia franciscana* in a benthic environment: Size and shape-dependent toxicity. *Mar. Pollut. Bull.* **2024**, *209*, 117270. [CrossRef] [PubMed]
8. Klimasz, M.; Grobelak, A. Effects of Microplastics on Selected Earthworm Species. *Toxics* **2025**, *13*, 201. [CrossRef] [PubMed]

toxics

MDPI

Article

Microplastics and Endocrine Disruptors in Typical Wastewater Treatment Plants in Megacity Shanghai

Yuxiao Tong [1], Manjun Xie [2], Hanwen Xv [1], Ruihua Sun [3], Qian Wang [1,*] and Juanying Li [1]

1 College of Oceanography and Ecological Science, Shanghai Ocean University, Shanghai 201306, China; tyx0061@163.com (Y.T.); m210501173@st.shou.edu.cn (H.X.); jyli@shou.edu.cn (J.L.)
2 Shanghai Haibin Sewage Treatment Plant, Pudong, Shanghai 201306, China; xiemanjunkeynes@163.com
3 Pudong Environmental Monitoring Station, Pudong, Shanghai 201306, China; rhsun2@hotmail.com
* Correspondence: q_wang@shou.edu.cn

Abstract: The fast development of China's urbanization has led to a notable release of emerging pollutants, including microplastics (MPs) and endocrine disruptors (EDCs). Generally, these pollutants enter the coastal environment through the discharge of wastewater treatment plants (WWTPs) and finally threaten the organisms in the receiving waterbody. The study investigated the environmental behavior of MPs and EDCs in two typical WWTPs in one of the megacities in China, Shanghai. The abundance of MPs in the influent ranged from 321 to 976 items/L. Four shapes (films, fragments, fibers, and microbead) were found, while fibers and films dominated. Transparent (31–63%) and white (20–47%) MPs were more frequently observed, while polyethylene terephthalate, cellulose, and cellophane were the main polymetric materials. The size of the MPs fell between 15.8 μm and 2220 μm, and the smaller one (<500 μm) dominated. The removal efficiencies of the two WWTPs for MPs ranged from 64% to 92%, and both WWTPs performed better for large pieces of MPs (>500 μm). For EDCs, total concentrations in the influent were detected, ranging from 113 to 2780 ng/L. Two groups, including phenolic estrogens (PEs) and steroid estrogens (SEs), were detected, and PEs, especially bisphenol A (BPA), were the predominant individuals among the studied EDCs. Specifically, PEs ranged from 82.8 to 2637 ng/L, while SEs ranged from 27.3 to 143 ng/L. The removal efficiencies of the WWTPs for EDCs varied (82.8–100%) as well, possibly due to the different treatment compartments and contamination load in the influent. Seasonal variations for both MPs and EDCs were observed. Specifically, concentrations of MPs and EDCs in WWTPs influent were higher in the wet season, as well as the removal efficiency. Furthermore, there was a correlation observed between the concentrations of MPs and EDCs, suggesting that MPs and EDCs may originate from the same source and that EDCs released by MPs cannot be ignored during treatment. Finally, the study evaluated the environmental risk of the effluents. MPs led to a minor risk (Level I), while EDCs might lead to an adverse impact on algae (RQs = 0.0014–0.024) and fish (RQs = 3.4–30.2). In summary, WWTPs received considerable amounts of MPs and EDCs. Although the WWTPs removed the contaminants efficiently, the environmental risk of the effluent needs to be noted.

Keywords: microplastics; EDCs; WWTPs; environmental risk

1. Introduction

In recent decades, the urbanization of China has been fast, and Shanghai, as one of the most important megacities, uses and releases tons of emerging contaminants (ECs) from its human activities [1]. ECs included current-use pesticides (CUPs), pharmaceuticals and personal care products (PPCPs), endocrine disruptors (EDCs), microplastics (MPs), etc. These ECs are treated by sewage treatment plants, discharged into the coastal environment, and accumulate, eventually causing negative effects on ecosystems (biodiversity, sex, or age structure of the population) and aquatic life (individual death, developmental disorders, dysgenesis, etc.) [2]. However, the WWTPs are not designed for ECs specifically. Generally, primary, secondary, and tertiary treatment processes are included in WWTPs, and they are effective in eliminating pollutants such as organic substances and nutrients. The ECs could be eliminated via sedimentation, adsorption, and biodegradation [3,4]. However, the removal efficiencies are not stable, depending on the types, loads, and characteristics of the contamination. WWTPs have been identified as a non-negligible point source of ECs in the environment due to their unstable removal and continuous discharge [2].

Among the ECs, MPs and EDCs are receiving more and more concerns due to their rising adverse environmental impacts. Take MPs, for example; China produced more than 1.0 billion tons of plastic materials in 2020 (National Bureau of Statistics of China, 2000–2021). Many of the waste plastics are disposed of disorganized and broken into MPs (<5 mm). MPs in cities are partially discharged into WWTPs and finally released into the freshwater or coastal environment [5,6]. Thus, WWTP is an important point source of MPs. As reported, WWTPs in another megacity, Shenzhen in south China, produced 70.6–302 tons of MPs into its connected coast annually [7]. Based on the information provided by Ren, et al. [7], WWTPs across the country discharge 734–3100 tons of MPs into surface water each year, of which 220 to 950 tons are discharged annually into the marine environment. Thus, it is important to confirm and improve the removal of MPs in WWTPs. Although the WWTPs can achieve acceptable removal of MPs, the removal rate was unstable, and the removal efficiencies in different compartments were unclear [8]. For example, MPs could be removed by coagulation/flocculation (primary treatment) with a ratio of 47–82% and by A^2O (secondary treatment) with a range of 47–82%, both of which fall within a wide range [9]. The removal rates highly depended on the contamination loads and composition patterns of MPs. Therefore, it is necessary to study the environmental behavior of MPs in and through the WWTPs in the megacity of Shanghai and evaluate the environmental risk of effluent that carries MPs.

Plastics and MPs have been confirmed as carriers for EDCs. In the polypropylene resin granules collected from the coastline in Japan, nonyl phenol (NP) was detected at a level of 0.13–16 ng/g [10]. During the aging process, plastics and MPs released a considerable amount of bisphenol A (BPA), and the aged plastics were identified as a continuous source of BPA [11]. Thus, the presence of MPs might be an indicator of the contamination of EDCs. It is necessary to study MPs and EDCs comprehensively to explain their environmental behavior. In addition to MP releases, industrial and medicinal procedures may potentially release EDCs as well. Specifically, steroid estrogens (SEs), such as estrone (E1), estradiol (E2), and estriol (E3), are used to regulate the development and reproduction function of human beings [12], while phenolic estrogens (PEs), such as BPA and NP, are important materials to produce surfactants, detergents, and paper products [13,14]. These EDCs were initially piped into the WWTPs and finally discharged into the receiving waterbody, e.g., the Yangtze River Estuary, in this study. The intensive discharge of EDCs and their metabolites into the environment might lead to hormonal dysfunction or reproductive abnormalities and perturb the stability of ecological systems [15]. However, as investigated by Xu, et al. [16], 12 EDCs were observed in the WWTP effluent in Hong Kong, in which the mean concentrations of E1 and NP were 5.25 ng/L and 4510 ng/L, respectively. Stasinakis, et al. [17] also reported a high level of BPA, which is of up to 5.76 μg/L. Similarly, the WWTPs are not specifically designed to eliminate EDCs, and the removal rates vary. The WWTPs in Hong Kong even reported elevated concentrations for some EDC individuals in their effluent [16]. Research on the removal of EDCs through WWTPs is essential for assessing the environmental risk of the effluent and improving treatment techniques, more specifically, for those in megacities that discharge significant amounts of ECs into the receiving water bodies.

Based on two typical WWTPs in Shanghai, the study evaluated the behavior and seasonal variation of MPs (sharp, color, polymers, and size) and EDCs (7 individuals) in and through different compartments in the WWTPs and confirmed removal efficiencies of different compartments. Then, the source of MPs and EDCs in the influent was identified. Subsequently, the relationship between the abundance of MPs and the concentration of

EDCs was analyzed using principal component analysis (PCA). The environmental risk of MPs and EDCs in the effluent was calculated using the ecological risk index and risk quotient methods. Finally, the study could provide support for treatment optimization in WWTPs, as well as pollution management of ECs.

2. Materials and Methods

2.1. Materials and Chemicals

The chemicals for MP sample preparation included hydrogen peroxide (H_2O_2, analytical grade) and sodium chloride (NaCl, analytical grade). They were purchased from Sinopharm Group Chemical Reagent Co., Ltd. (Shanghai, China). For the analysis of EDCs (7 in total, Tables S3 and S4), their native standard solutions and surrogates (CRM) were needed. The standard substances, such as SEs (E1, E2, EE2, and E3), PEs (BPA and NP), and the required isotope-labeled surrogates were obtained from Dr. Ehrenstorfer GmbH (Augsburg, Germany). Solid-phase extraction (SPE) cartridges (Oasis HLB, 500 mg, 6 mL) to remove the interferences in the samples were provided by Waters Corporation (Milford, USA). The organic solvents (HPLC grade) were provided by Adamas (Shanghai, China), while the deionized water was produced by the Milli-Q®system in the laboratory (18.2 MΩ·cm).

2.2. Field Sampling

The study focused on two typical wastewater plants in Shanghai, namely WWTP-A (2.8 million tons/day) and WWTP-B (400,000 tons/day). WWTP-A is the largest wastewater treatment facility in Shanghai, and WWTP-B includes advanced tertiary treatment processes, like ozone oxidation, which is not usual in traditional WWTPs. The main treatment processes of WWTP-A and WWTP-B are given in Figure 1. Water samples (n = 3, V = 1 L) were collected at the outlet of each treatment compartment in wet (July 2023) and dry (March 2023) seasons, respectively. Considering the hydraulic retention time (HRT) of the specific WWTP, each influent and effluent was collected and mixed every 24 h. Samples were kept in an amber bottle that had been rinsed with a solvent and pure water. Afterward, 4 M sulfuric acid (1/2000, V/V) was spiked to reduce biodegradation. The samples were transported to the laboratory, kept at 4 °C, and processed within 48 h.

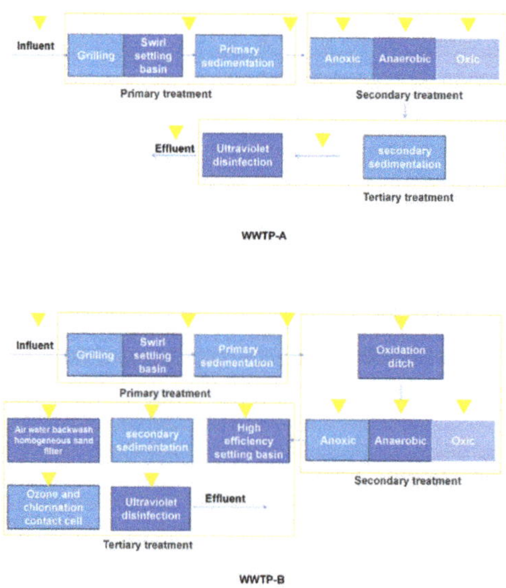

Figure 1. Treatment Compartments of WWTPs A and B. The yellow arrows represent sampling locations in the WWTPs.

2.3. Sample Preparation

Water samples were filtered through glass fiber filters (Whatman GF/F, 0.45 μm, Maidstone, UK). The suspended solid and MPs that remained on the filters were reserved for further treatment, and the filtrated water was processed with SPE cartridges for EDCs analysis.

The MPs that remained on the filters were washed by H_2O_2 into bakers, which were initially rinsed with pure water. Subsequently, the mixtures were placed into a shaker at a condition of 65 °C and 80 rpm for 72 h. Finally, the mixtures were processed again using membranes with a pore size of 5 μm before further analysis. More details have been provided by Su, et al. [18]

For EDCs, the filtrates were spiked with 0.2 g Na_2EDTA and extracted by SPE with Oasis HLB cartridges (500 mg and 6 mL, Waters, Milford, MA, USA). The SPE cartridges were preconditioned with 10 mL methanol and 10 mL Milli-Q water. Then, water samples were loaded onto the cartridges at a flow rate of 10 mL/min. Afterward, the cartridges were rinsed with 10 mL of Milli-Q. The EDCs retained on the cartridges were eluted with 8 mL methanol and 8 mL methanol: Acetone (1:1, *v*/*v*), then the eluates were condensed to almost dry using a nitrogen evaporator and re-dissolved in 300 μL acetonitrile: water (3:7, *v*/*v*). The final extract was transferred to a 1 mL amber vial and stored at −18 °C until ultra-high-performance liquid chromatography-tandem mass spectrometry (UHPLC-MS/MS) analysis.

2.4. Instrumental Analysis

MPs were observed and shot using a stereo microscope and a photographic system (Nikon SMZ25, Nikon Corporation, Yokohama, Japan). Cooperating with the photographic system, the software (NIS-Elements D 4.50.00, Nikon Corporation, Yokohama, Japan) was used to record information about the physical properties of plastics, such as shape, color, and size. The imageJ (1.5) software was applied to label the scale of the photographic process. Based on the ratio of the pixel and scale, the size of MPs could be determined. Then, the polymeric component of MPs was analyzed using a Fourier transform infrared micro spectrometer (Nicolet iN 10 MX, Thermo Fisher, Waltham, MA, USA) with a transmission mode-MCT detector. More details of sample preparation for MPs could be found in the previous study [19].

Identification and quantification of EDCs were conducted using Ultra-High-Performance Liquid Chromatography-tandem mass spectrometry (UHPLC-MS/MS) equipped with an electrospray ionization source (Agilent 1290 Infinity UHPLC system and Agilent 6460 triple quadrupole MS; Agilent, Palo Alto, CA, USA) in multiple-reaction monitoring (MRM) mode. The analytes were separated on an Agilent Zorbax RR Eclipse Plus C18 column (95 Å pore size, 3.5 μm particle size, 2.1 mm inner diameter, and 150 mm length). The column temperature was set at 40 °C. The analysis EDCs was carried out using 0.1% ammonium hydroxide/H_2O as eluent A and acetonitrile as eluent B at a flow rate of 0.3 mL/min; the injection volume was 10 μL. The ion source parameters for both positive (ESI+) and negative (ESI−) modes were selected as follows: gas temperature 300 °C, gas flow 7 L/min (antibiotics) and 10 L/min (EDCs), nebulizer 45 psi, sheath gas temperature 350 °C, sheath gas flow 11 L/min, and the capillary was 3500 V. Nitrogen gas was used as the collision gas. More details on elution gradients and precursors/products after MRM reaction are given in Tables S2 and S3.

2.5. Quality Control

To ensure the data quality of MPs, all laboratory testing is conducted in a specialized facility with restricted access. Blank samples are prepared using a blank filter with 1 L of pure water passing through. To minimize the contamination, all operators were asked to wear cotton lab coats to minimize the introduction of synthetic fibers. All the glass wares were rinsed with pure water. Blanks for EDCs were made for each of the ten samples. The MDL was estimated using a signal-to-noise ratio of 3 (Table S5). For the recovery study, water samples were spiked with standards at a concentration of 100 ng/L, and recovery ratios were also given in Table S5.

2.6. Risk Assessment

Environmental risk of MPs in the influent and effluent water was evaluated using the ecological risk index method, which was developed by Hakanson [20] and improved by Peng, et al. [21]. The calculations are given as follows:

$$C_f^i = \frac{C^i}{C_n^i} \tag{1}$$

$$T_r^i = \frac{P_i}{C^i} \times S_i \tag{2}$$

$$E_n^i = T_n^i \times C_f^i \tag{3}$$

$$RI = \sum\nolimits_{i=1}^{n} E_n^i \tag{4}$$

C_f^i is the pollution index which could be yielded using the measured abundance (C^i) divided by the criterion reference value (C_n^i), which in this case refers to the safe concentration of MPs in surface water (6650 items/m^3) estimated by Everaert, et al. [22]. T_n^i is the ecotoxicity response factor representing the toxicity and bio-sensitivity of the MPs. According to the study of Lithner, et al. [23], P_i is the abundance of specific MP polymer (i), and S_i is the hazard index of polymer (i) (Table S6). E_r^i is the potential ecological risk index, which is determined based on T_r^i and C_f^i of each polymer. RI is the total ecological risk index of all types of MP polymers. As given in Table 1, ecological risk is classified as level I, II, III, and IV according to different RI values.

Table 1. Risk-level criteria for microplastic pollution.

Potential Ecological Risk Factor	Ecological Risk level
<10	I
10–100	II
100–1000	III
>1000	IV

The ecological risk caused by EDCs was quantified based on the risk quotient method (RQ), which could be calculated by the measured concentration (MEC) divided by the predicted no-observable concentration (PNEC) [24]. The total RQ values (ΣRQ) of all the studied EDCs could be determined by summing up all the RQ values of individuals. If ΣRQ values are lower than 0.01, no risk could be identified. Low risk could be identified if ΣRQ values fall between 0.01 and 0.1. If ΣRQ values fall between 0.1 and 1.0, the risk is at an intermediate level. If ΣRQ values > 10, the risk is at a high level [25]. The PENC values used in the environmental risk assessment are shown in Table S7. The effluent in this study was discharged into the Yangtze River Estuary generally, and PNEC values for algae and fish were selected. This evaluation is important to quantify the impact of the EDCs on the primary productivity and ecosystem structure in the receiving waterbody.

3. Result and Discussion

3.1. Behavior of MPs in and through Two Typical WWTPs in Megacity

3.1.1. Contamination Patterns of MPs in the Influent and Source Analysis

The abundances of MPs in the influent from the two WWTPs were different, and a significant seasonal variation was also noted. Specifically, the abundances were 93 items/L in WWTP-A and 163 items/L in WWTP-B in the wet season, while 54 items/L WWTP-A and 140 items/L in WWTP-B in the dry season (Table S8). The observed abundance in this study was comparable to those found in India (42–150 items/L) [26]. Furthermore, the abundance in the wet season was significantly higher than that in the dry season. Similar seasonal variation was found in previous studies [27,28]. As previously reported for ions and solids, increased concentration of MPs in influent during the wet season (July) could be explained by sun irradiation, which enhances water evaporation [29]. The higher temperature in the wet season also leads to more plastic consumption and washing activities [8,30]. More water evaporation, plastic consumption, and washing activities (major sources of fibers) comprehensively increased the MPs abundance in the influent.

Four sharps of MPs were detected, namely film, fragment, fiber, and microbeads (Figure 2a). In WWTP-A, fiber, which is primarily from domestic washing, was dominated with a ratio of 50–60%. This finding was the same as previous studies [31]. During washing activities, superfine fibers are flushed from the synthetic clothing into the sewage system [32]. As estimated, approximately 100 fibers in 1 L of laundry wastewater enter WWTPs [33]. Insufficient removal of fibers occurs due to their smaller size and lower density [6] and the residues are introduced into the receiving environment finally. The second dominant sharp was film, generally from the plastics used in packaging [34]. In WWTP-B, the sharp film dominated, followed by fiber, indicating a different source of MPs in the influent. The abundance of microbeads that were used as ingredients in toothpaste and personal care products exhibited the lowest level, indicating the positive effect of the prohibition of microbeads in personal care products [35].

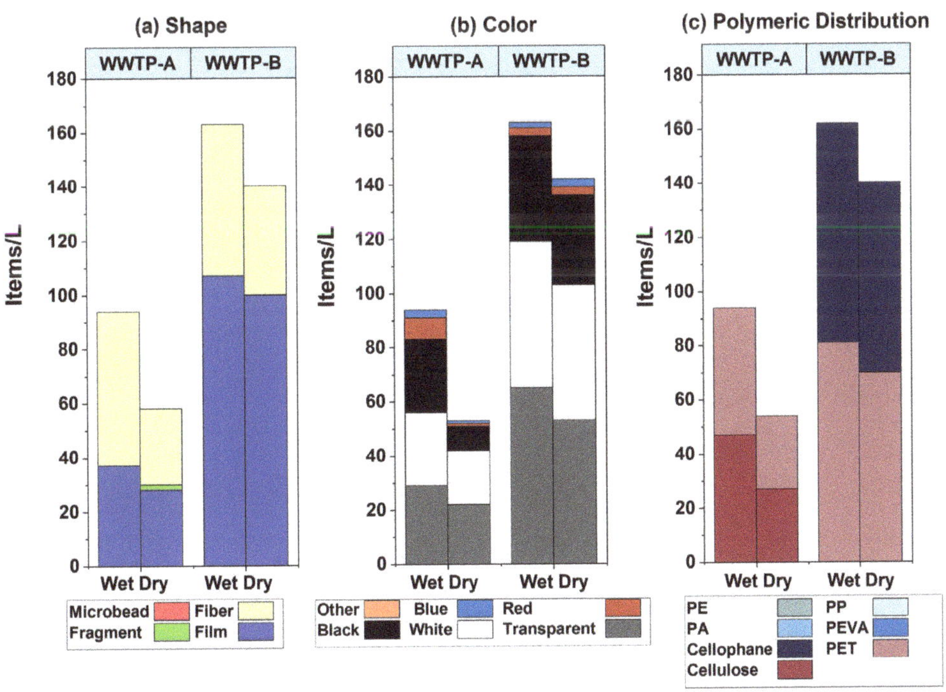

Figure 2. Shape (**a**), color (**b**) and polymeric distribution (**c**) of MPs in the influent samples from the two WWTPs. PE: polyethylene, PP: polypropylene, PA: polyamide, PEVA: polyethylene vinyl acetate, PET: polyethylene glycol terephthalate.

Transparent (31–63%) and white (20–47%) were the most common colors, followed by black (11–40%) and other colors (Figure 2b). Transparent and white MPs were primarily from packaging materials, such as bottles and plastic tableware. The color of MPs determines the hazardous potential to the biota. Transparent and white MPs are easily ingested by zooplankton, fish, and other species, causing environmental risks [36]. Therefore, it is crucial to characterize the color of MPs and assess their potential risks.

Polymeric distribution of the MPs was determined. Polyethylene glycol terephthalate (PET) (50%) and cellulose (50%) dominated WWTP-A, while PET (50%) and cellophane (50%) dominated WWTP-B (Figure 2c). No significant seasonal variation in polymeric distribution was observed. PET was the major polymer in all the samples, mainly due to its extensive application in synthetic clothing and food packaging [37]. Cellulose has been used widely in industries such as mechanical engineering and biopharmaceutical [38], indicating a wastewater source from these industries in WWTP-A. Cellophane has been confirmed to be generated from washing activities, suggesting a domestic source [39]. Different polymeric distributions indicate a different water source for the two WWTPs.

3.1.2. Behavior of MPs through the WWTP

The behavior of MPs through the WWTP in different seasons was studied. In general, the abundance of MPs reduced. The removal rates in WWTP-A in the wet and dry seasons were 82.1% and 63.9%, while those in WWTP-B were 92.1% and 88.6% (Table S6). Removal rates in the wet season were higher than those in the dry season.

Figure 3 shows the changes in size and polymeric distribution of MPs in different compartments. MP size determined the removal efficiency significantly. Generally, the MPs with sizes >500 μm were reduced significantly from influent to secondary treatment (Table S9) in both WWTPs. However, it is noted that the abundance of MPs with sizes 500–1000 μm and ≥1000 increased with a ratio of 27 (Table S9) in the dry season in WWTP-A. The last step in primary treatment in WWTP-A is a sedimentation tank, which may release the settled or adsorbed MPs back into the water phase [40]. Also, there is a slight increase after the secondary sedimentation of the tertiary treatment, possibly due to the resuspension from the sludge in this process [40]. For those MPs between 100 and 500 μm, the abundance did not change significantly in WWTP-A, while it reduced notably in WWTP-B. For the MPs even smaller (<100 μm), the primary treatment did not perform well, while the tertiary showed better efficiency. Finally, in the effluent, the MPs within the 100–500 μm range were more frequently observed, followed by those smaller than 100 μm. The smaller pieces float more easily, which impedes their settling into the sludge [41].

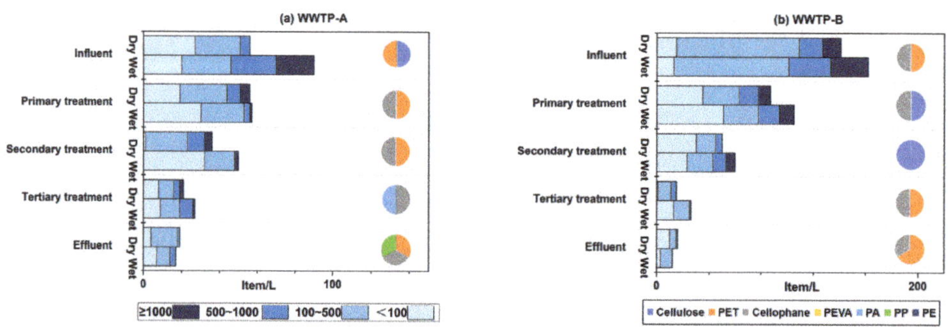

Figure 3. Size and percentage polymeric distribution of MPs in different compartments in the WWTP-A (**a**) and WWTP-B (**b**).

PET was removed sufficiently for the most abundant polymer. Especially in WWTP-B, PET was removed at a rate of 100% (Table S10). The high reduction of PET in the WWTP was likely due to its high density (1.38 g/cm^3) [28,42]. However, PET concentration was elevated after the tertiary treatment compartment. The release of PET from sludge or adsorption materials may explain this increase. Furthermore, the polymer polypropylene

(PP), which was not detected in the influent in both WWTPs, was observed in the effluent, indicating a considerable release from the PP pipelines used in the WWTPs system [43]. Conclusively, WWTPs performed well for MPs, especially for those with large size and high density.

3.2. EDCs in the WWTPs

3.2.1. EDCs in the Influent and Source Diagnosis

In this study, seven EDCs were analyzed in the samples collected in March (dry season) as well as in July (wet season) from the WWTPs. The results are shown in Figure 4a,b, and Table S11, where ∑EDCs concentrations in the influent from WWTP-A were found to be 485.46 ± 10.98 ng/L in the wet season and 381.85 ± 0.54 ng/L in the dry season, while those from WWTP-B were 2780.45 ± 59.03 ng/L and 113.85 ± 0.61 ng/L, respectively. Five target EDCs (BPA, NP, E1, E2, and E3) were detected in the influent water from both WWTPs, and the dominant individuals were BPA and NP. BPA is generally from paper mills, fine chemical facilities, and plastics production, while NP comes from the production of detergent products and textile auxiliaries [14,44]. The presence of BPA and NP also suggested that industrial wastewater is a significant influent source of the two WWTPs.

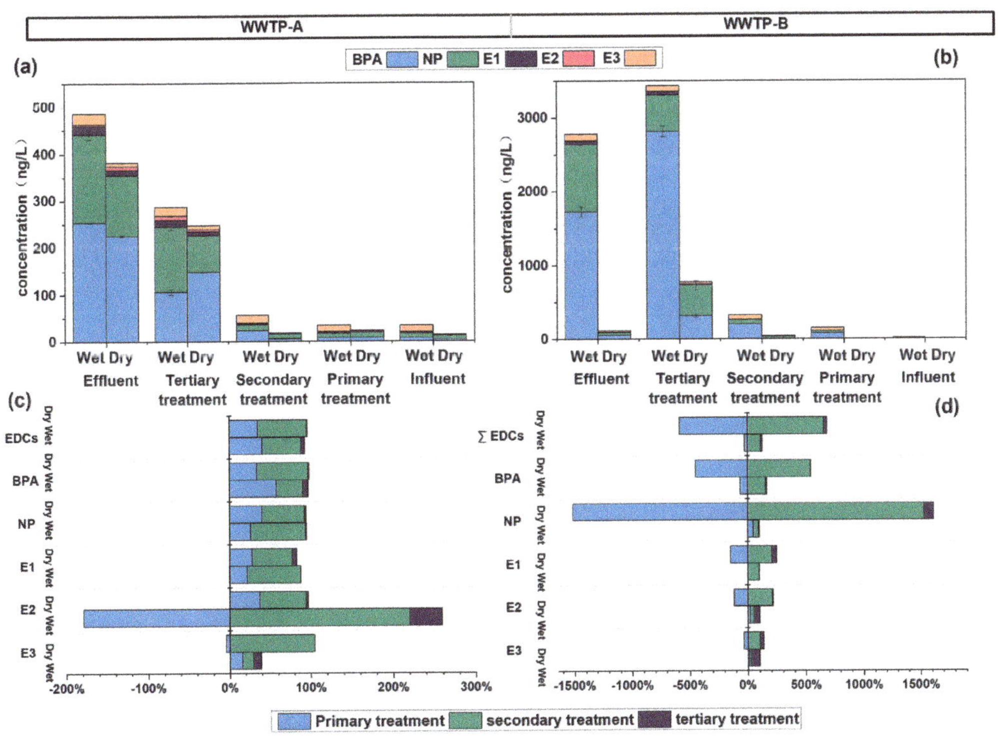

Figure 4. EDCs in different treatment compartments (**a**,**b**) and the removal rates (**c**,**d**).

3.2.2. The Removal Efficiencies of EDCs through Different Treatment Compartments

After the treatments, the concentration was reduced significantly (Figure 4a,b, and Table 2). Specifically, the removal rates of total EDCs were 93.22% and 96.52% in the wet and dry seasons in WWTP-A, while 99.64% and 98.98% in WWTP-B, respectively.

Table 2. The removal rates of the EDCs through the two typical WWTPs in different seasons.

		Wet				Dry			
		Primary Treatment	Secondary Treatment	Tertiary Treatment	Total Removal Rate	Primary Treatment	Secondary Treatment	Tertiary Treatment	Total Removal Rate
WWTP-A	BPA	58.25%	77.97%	69.90%	97.23%	34.17%	96.12%	34.40%	98.32%
	NP	26.26%	91.46%	24.10%	95.22%	40.29%	87.84%	22.26%	94.36%
	E1	21.74%	84.41%	−3.31%	87.39%	27.83%	68.44%	24.26%	82.75%
	E2	−178.99%	78.91%	65.90%	79.93%	37.33%	91.01%	44.88%	96.89%
	E3	15.50%	16.47%	13.60%	39.02%	−4.53%	100.00%	100.00%	100.00%
	∑ EDCs	40.91%	80.67%	40.62%	93.22%	35.30%	92.63%	26.92%	96.52%
WWTP-B	BPA	63.25%	92.98%	97.86%	99.76%	−447.56%	98.95%	76.17%	98.63%
	NP	46.13%	88.03%	91.03%	99.42%	−1513.88%	94.80%	99.45%	99.54%
	E1	2.67%	96.44%	64.68%	98.77%	−146.65%	81.77%	90.55%	95.75%
	E2	20.73%	40.23%	100.00%	100.00%	−115.30%	96.64%	100.00%	100.00%
	E3	13.33%	18.24%	100.00%	100.00%	−33.71%	77.80%	100.00%	100.00%
	∑ EDCs	−23.54%	90.57%	96.89%	99.64%	−585.78%	95.71%	96.52%	98.98%

In terms of the removal efficiencies in different treatment compartments, the primary treatment in WWTP-A eliminated total EDCs with a rate ranging from 35.3% (dry) to 40.9% (wet), and the secondary process removed EDCs with a percentage of 80.7% (wet)–82.6% (dry) (Table 2). Tertiary treatment, as well as disinfection before discharge, did not significantly remove EDCs. Unlike WWTP-A, primary treatment in WWTP-B elevated EDCs concentration in both studied seasons (Figure 4a,b, Tables 2 and S11), which may be related to the re-suspended of these EDCs from the sludge. The secondary and tertiary treatments (ozone oxidation included) performed well, with their removal rates higher than 90%. Filtration, sedimentation, and adsorption are basic processes in the primary treatments to remove contaminants; biodegradation and bio-absorption are major processes in the secondary treatments, and in tertiary treatments, the contaminants could be eliminated through advanced oxidation or adsorption processes. In conclusion, it is demonstrated that the principal process for removing EDCs from wastewater is biochemical degradation during secondary treatment. Advanced oxidation is another effective process, as confirmed by WWTP-B.

In addition, there were large differences in the removal efficiency of the five detected EDC individuals. As shown in Figure 4c, in WWTP-A, removal variations of phenolic estrogen in different seasons were not significant, while those of steroid hormones in different seasons were significantly non-negligible. The performance in the wet season is better than in the dry season. As concluded in the last paragraph, the principal processes for removing EDCs from wastewater are biochemical degradation during secondary treatment, and higher temperature leads to enhanced biodegradation [45]. Also, E2 and E3 were released during the primary treatment. This may be due to the reconversion of inactive conjugated steroidal estrogens to active forms, ultimately leading to an increase in their concentrations [46]. In the tertiary treatment, the EDCs were removed but with a low removal efficiency (Table 2). Our result was comparable to previous studies in that micro-activities removed EDCs rather than sedimentation [47].

EDC individuals exhibited different behaviors in WWTP-B. Concentrations in WWTP-B were magnitudes higher in the wet season than those in the dry season. As plotted in Figure 4d, there were significant releases of BPA and NP after the primary treatment. Aeration processes in the primary treatment may lead to a release of BPA and NP from the sludge [45]. E1, E2, and E3 increased as well due to the uncoupling of the sulfate and glucuronide affixes of EDCs during the treatments [48]. This result emphasized that conjugate compounds play a non-negligible role in the removal of EDCs and deserve further exploration. Although there were releases from the primary treatment, the secondary treatment in WWTP-B showed a good removal for all five monomers. The phenolic estrogens (88.03% to 98.85%) were removed more efficiently than steroidal estrogens (18.24% to 96.64%) (Table 2). In summary, although some monomers presented an elevated trend in certain treatment compartments, both wastewater plants can remove more than 80% of the EDCs. The compartment (secondary treatment) based on biodegradation played a more important role.

3.3. Relationship between EDCs and MPs

To further confirm the possibility of MPs and EDCs as each other's indicators, a principal component analysis (PCA) was applied, and the result is given in Figure 5. As indicated, the concentration of EDCs and MPs with small size (<100 μm) was significantly correlated, indicating a potential of these MPs carrying or releasing the EDCs [49]. In a previous study on the environmental behavior of plastic additives [50], it was demonstrated that the presence of BPA in the water column is highly related to the migration of additives from plastic containers. Additionally, EDC levels negatively correlated to the abundance of PP, PA, and PE. These components of MPs possibly absorbed and accumulated EDCs on their surface due to the homogeneous structure and strong hydrophobicity, finally reducing EDC concentration in the dissolved phase. In summary, the correlation between EDCs and MPs suggested that the size and composition of MPs may affect the concentration of EDCs. More research regarding MPs as a contamination indicator for EDCs should be carried out in the future.

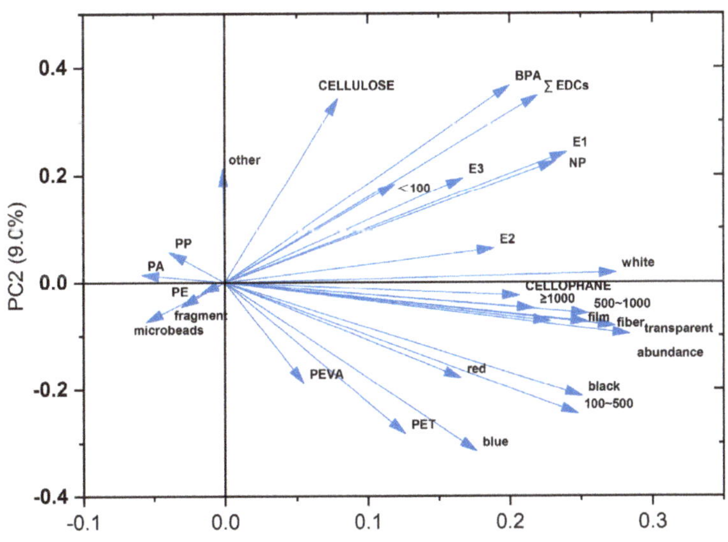

Figure 5. PCA analysis based on MP abundance, their properties, and EDC concentrations.

3.4. Risk Reduction by the WWTPs

Compared to the safe value (6650 items/m^3) proposed by Everaert, et al. [22], the abundances in the influent (53,500–162,667 items/m^3) and effluent (12,833–22,677 items/m^3) were significantly higher than the safe concentration. Although the MPs were removed efficiently, possible risk still could be initiated by the effluent. Comparatively, the MPs abundances in both influent (171,000 items/m^3) and effluent (12,800 items/m^3) from a WWTP from Spain were higher than the safe concentration [51]. Similar results were found in another WWTP in a big city, Xi'an, in China [52]. To evaluate the environmental risk more accurately, the ecological risk index of the WWTP wastewater was determined (Table 3). As given in the table, the wastewater would induce a risk at level II without WWTP treatments. After a series of treatments, the risk index values were reduced to a range of 1.89 to 4.28, which were at an I level. From the influent (RI = 8.1–24.5) to effluent (RI = 1.25–4.28), WWTPs significantly reduced the environmental risk initiated by MPs. It is noted that the risk value was not considered for cellulose and cellophane due to their limited hazardous data. The RI values were only calculated based on the PE, PA, PP, PEVA, and PET, leading to a possible underestimation.

Table 3. MPs risk evaluation in the influent and effluent in the WWTPs.

WWTPs	Sampling Season	Influent RI	Risk Level of Influent	Effluent RI	Risk Level of Effluent	Risk Reduction Rate (%)
WWTP-A	Wet	14.0	II	1.89	I	86.51%
	Dry	8.1	II	1.25	I	84.42%
WWTP-B	Wet	24.5	II	3.43	I	85.97%
	Dry	21.1	II	4.28	I	79.68%

The risk of EDCs was assessed based on the risk quotient method. As given in Table 4, both influent (RQ = 0.106–3.08) and effluent (RQ = 0.0014–0.024) pose limited risk to algae. The reason is that the EDCs primarily target endocrine systems, which algae does not have. After WWTPs, the risk of EDCs for algae reduced from intermediate to low. Algae is an important primary producer in an ecosystem. The removal of the EDCs reduced the environmental pressure on algae and protected the stability of the ecosystem. For fish, the risk of EDCs from WWTPs could not be ignored. Although the risk reduction rates (83.9–98.8%) were high, the residual levels of EDCs still trigger high risk to fish. Specifically, the RQ values of effluent ranged from 3.40 to 30.1, and the risk was primarily induced by E1 and BPA. The receiving water of the two WWTPs is an important waterbody for rare or endangered species, such as Chinese sturgeon and estuarine taper tail anchovy. Similarly, previous studies found environmental threats due to EDCs (BPA, E1, and E2) in the effluent from WWTPs in Europe [53,54].

Table 4. EDCs risk evaluation in the influent and effluent in the WWTPs.

			Risk to Algae			
WWTPs	**Sampling Season**	**Influent RQ**	**Influent Risk Level**	**Effluent RQ**	**Effluent Risk Level**	**Risk Reduction Rate (%)**
WWTP-A	Wet	0.547	Intermediate	0.024	Low	95.7%
	Dry	0.428	Intermediate	0.0175	Low	95.9%
WWTP-B	Wet	3.08	High	0.0126	Low	99.6%
	Dry	0.106	Intermediate	0.0014	Low	98.7%
			Risk to fish			
WWTP-A	Wet	239	High	30.1	high	87.4%
	Dry	160	High	25.8	high	83.9%
WWTP-B	Wet	648	High	7.54	high	98.8%
	Dry	87.8	High	3.40	high	96.1%

Thus, technical improvement is needed in WWTPs, especially for EDCs. Advanced oxidation is an effective option for the removal of EDCs. Some studies found that advanced oxidation using ozone could remove more than 90% of the residual EDCs [55–57], and WWTPs coupled with wetlands may further reduce the environmental risk of the effluent [58–60].

4. Conclusions

This study confirmed that the influence of the two WWTPs in the megacity of Shanghai was from both domestic and industrial wastewater. The abundance of MPs was higher in the wet season due to the higher temperature, increased plastic consumption, and washing activities. The removal of MPs in the WWTPs was good, and the wet season performed better than the dry season. MPs of large size and density were more easily removed by sedimentation, while those with small size and density were more difficult to remove. For EDCs, the removal was good as well, ranging from 93.2% to 99.6%. However, the removals of different EDC individuals varied considerably in different seasons and WWTPs. MPs could also be considered an indicator of EDC contamination in the WWTPs. MPs possibly released EDCs during the water treatment processes. Finally, the WWTPs cut off the environmental risk of wastewater significantly and reduced the environmental pressure on the receiving water bodies. The environmental risk caused by MPs in the effluent is low, but EDCs still cause non-negligible negative environmental impacts on fish and ultimately threaten the stability of the ecosystem. Technical improvement is needed in WWTPs, especially for EDCs. Advanced oxidation techniques, such as ozone oxidation, should be applied. WWTP coupled with wetlands is another alternative option. Also, the reuse of effluent water in a particular industry or agriculture instead of directly discharging it should be considered.

Supplementary Materials: The following supporting information can be downloaded at: https://www.mdpi.com/article/10.3390/toxics12050345/s1, Table S1: List of abbreviations; Table S2: Instrumental parameters for the target compounds; Table S3: Mass spectrum monitoring conditions of target compounds and internal standards; Table S4: Physicochemical properties of the target contaminants; Table S5: The recoveries (%), method detection limits (MDLs), and limits of quantification (LOQs) of the contaminants; Table S6: Ranking of plastic polymers based on hazard classifications of monomers; Table S7: PNECs values used in the environmental risk assessment for EDCs; Table S8: PAbundance of MPs with different sizes in different compartments; Table S9: Removal efficiencies of MPs with different sizes of each treatment compartment; Table S10: Removal rates of different polymers through the WWTPs; Table S11: Concentrations of different EDCs in different compartments. References [61–65] are cited in the supplementary materials

Author Contributions: Conceptualization, Q.W. and J.L.; methodology, Y.T. and H.X.; software, Y.T. and H.X.; validation, Q.W. and J.L.; formal analysis, Y.T. and H.X.; investigation, M.X. and R.S.; resources, Q.W. and J.L.; data curation, Y.T. and H.X.; writing—original draft preparation, Y.T.; writing—review and editing, Q.W.; visualization, Y.T.; supervision, Q.W.; project administration, Q.W.; funding acquisition, Q.W. and J.L. All authors have read and agreed to the published version of the manuscript.

Funding: This research was funded by [the Science, Technology and Economy Commission of Pudong New Area, Shanghai] grant number [Grant No. PKJ2021-C04] and [Technology Innovation Center for Land Spatial Eco-restoration in Metropolitan Area, Ministry of Natural Resources] and [the Fundamental Research Funds for the Central Universities] grant number [CXZX2023A01] and [the program of opening ceremony to select the best candidates of the Key Laboratory of Marine Ecological Monitoring and Restoration Technology] grant number [MEMRT2024JBGS01].

Institutional Review Board Statement: Not applicable.

Informed Consent Statement: Not applicable.

Data Availability Statement: The data presented in this study are available on request from the corresponding author.

Conflicts of Interest: The authors declare no conflict of interest.

References

1. Rodriguez-Mozaz, S.; Lopez de Alda, M.J.; Barceló, D. Advantages and limitations of on-line solid phase extraction coupled to liquid chromatography–mass spectrometry technologies versus biosensors for monitoring of emerging contaminants in water. *J. Chromatogr. A* **2007**, *1152*, 97–115. [CrossRef]
2. Al Aukidy, M.; Verlicchi, P.; Jelic, A.; Petrovic, M.; Barcelò, D. Monitoring release of pharmaceutical compounds: Occurrence and environmental risk assessment of two WWTP effluents and their receiving bodies in the Po Valley, Italy. *Sci. Total Environ.* **2012**, *438*, 15–25. [CrossRef]
3. Rehman, M.U.; Nisar, B.; Mohd Yatoo, A.; Sehar, N.; Tomar, R.; Tariq, L.; Ali, S.; Ali, A.; Mudasir Rashid, S.; Bilal Ahmad, S.; et al. After effects of Pharmaceuticals and Personal Care Products (PPCPs) on the biosphere and their counteractive ways. *Sep. Purif. Technol.* **2024**, *342*, 126921. [CrossRef]
4. Stasinakis, A.S.; Thomaidis, N.S.; Arvaniti, O.S.; Asimakopoulos, A.G.; Samaras, V.G.; Ajibola, A.; Mamais, D.; Lekkas, T.D. Contribution of primary and secondary treatment on the removal of benzothiazoles, benzotriazoles, endocrine disruptors, pharmaceuticals and perfluorinated compounds in a sewage treatment plant. *Sci. Total Environ.* **2013**, *463–464*, 1067–1075. [CrossRef]
5. Murphy, F.; Ewins, C.; Carbonnier, F.; Quinn, B. Wastewater Treatment Works (WwTW) as a Source of Microplastics in the Aquatic Environment. *Environ. Sci. Technol.* **2016**, *50*, 5800–5808. [CrossRef]
6. Talvitie, J.; Mikola, A.; Koistinen, A.; Setälä, O. Solutions to microplastic pollution—Removal of microplastics from wastewater effluent with advanced wastewater treatment technologies. *Water Res.* **2017**, *123*, 401–407. [CrossRef]
7. Ren, S.-Y.; Sun, Q.; Xia, S.-Y.; Tong, D.; Ni, H.-G. Microplastics in wastewater treatment plants and their contributions to surface water and farmland pollution in China. *Chemosphere* **2023**, *343*, 140239. [CrossRef]
8. Ziajahromi, S.; Neale, P.A.; Leusch, F.D.L. Wastewater treatment plant effluent as a source of microplastics: Review of the fate, chemical interactions and potential risks to aquatic organisms. *Water Sci. Technol.* **2016**, *74*, 2253–2269. [CrossRef]
9. Sol, D.; Laca, A.; Laca, A.; Díaz, M. Approaching the environmental problem of microplastics: Importance of WWTP treatments. *Sci. Total Environ.* **2020**, *740*, 140016. [CrossRef]
10. Mato, Y.; Isobe, T.; Takada, H.; Kanehiro, H.; Ohtake, C.; Kaminuma, T. Plastic Resin Pellets as a Transport Medium for Toxic Chemicals in the Marine Environment. *Environ. Sci. Technol.* **2001**, *35*, 318–324. [CrossRef]
11. Sun, P.; Liu, X.; Zhang, M.; Li, Z.; Cao, C.; Shi, H.; Yang, Y.; Zhao, Y. Sorption and leaching behaviors between aged MPs and BPA in water: The role of BPA binding modes within plastic matrix. *Water Res.* **2021**, *195*, 116956. [CrossRef]
12. Wee, S.Y.; Aris, A.Z. Endocrine disrupting compounds in drinking water supply system and human health risk implication. *Environ. Int.* **2017**, *106*, 207–233. [CrossRef]
13. Kumar, P.; Shimali; Chamoli, S.; Khondakar, K.R. Advances in optical and electrochemical sensing of bisphenol a (BPA) utilizing microfluidic Technology: A mini perspective. *Methods* **2023**, *220*, 69–78. [CrossRef]
14. Devi, T.; Saleh, N.M.; Kamarudin, N.H.N.; Roslan, N.J.; Jalil, R.; Hamid, H.A. Efficient adsorption of organic pollutants phthalates and bisphenol A (BPA) utilizing magnetite functionalized covalent organic frameworks (MCOFs): A promising future material for industrial applications. *Ecotoxicol. Environ. Saf.* **2023**, *268*, 115706. [CrossRef]
15. Shi, W.-J.; Jiang, Y.-X.; Huang, G.-Y.; Zhao, J.-L.; Zhang, J.-N.; Liu, Y.-S.; Xie, L.-T.; Ying, G.-G. Dydrogesterone Causes Male Bias and Accelerates Sperm Maturation in Zebrafish (Danio rerio). *Environ. Sci. Technol.* **2018**, *52*, 8903–8911. [CrossRef]
16. Xu, E.G.B.; Liu, S.; Ying, G.-G.; Zheng, G.J.S.; Lee, J.H.W.; Leung, K.M.Y. The occurrence and ecological risks of endocrine disrupting chemicals in sewage effluents from three different sewage treatment plants, and in natural seawater from a marine reserve of Hong Kong. *Mar. Pollut. Bull.* **2014**, *85*, 352–362. [CrossRef]
17. Stasinakis, A.S.; Gatidou, G.; Mamais, D.; Thomaidis, N.S.; Lekkas, T.D. Occurrence and fate of endocrine disrupters in Greek sewage treatment plants. *Water Res.* **2008**, *42*, 1796–1804. [CrossRef]
18. Su, L.; Xue, Y.; Li, L.; Yang, D.; Kolandhasamy, P.; Li, D.; Shi, H. Microplastics in Taihu Lake, China. *Environ. Pollut.* **2016**, *216*, 711–719. [CrossRef]
19. Su, L.; Nan, B.; Craig, N.J.; Pettigrove, V. Temporal and spatial variations of microplastics in roadside dust from rural and urban Victoria, Australia: Implications for diffuse pollution. *Chemosphere* **2020**, *252*, 126567. [CrossRef]
20. Hakanson, L. An ecological risk index for aquatic pollution control.a sedimentological approach. *Water Res.* **1980**, *14*, 975–1001. [CrossRef]
21. Peng, G.; Xu, P.; Zhu, B.; Bai, M.; Li, D. Microplastics in freshwater river sediments in Shanghai, China: A case study of risk assessment in mega-cities. *Environ. Pollut.* **2018**, *234*, 448–456. [CrossRef]
22. Everaert, G.; Van Cauwenberghe, L.; De Rijcke, M.; Koelmans, A.A.; Mees, J.; Vandegehuchte, M.; Janssen, C.R. Risk assessment of microplastics in the ocean: Modelling approach and first conclusions. *Environ. Pollut.* **2018**, *242*, 1930–1938. [CrossRef]
23. Lithner, D.; Larsson, Å.; Dave, G. Environmental and health hazard ranking and assessment of plastic polymers based on chemical composition. *Sci. Total Environ.* **2011**, *409*, 3309–3324. [CrossRef]
24. Sharma, B.M.; Bečanová, J.; Scheringer, M.; Sharma, A.; Bharat, G.K.; Whitehead, P.G.; Klánová, J.; Nizzetto, L. Health and ecological risk assessment of emerging contaminants (pharmaceuticals, personal care products, and artificial sweeteners) in surface and groundwater (drinking water) in the Ganges River Basin, India. *Sci. Total Environ.* **2019**, *646*, 1459–1467. [CrossRef]
25. Hernando, M.D.; Mezcua, M.; Fernández-Alba, A.R.; Barceló, D. Environmental risk assessment of pharmaceutical residues in wastewater effluents, surface waters and sediments. *Talanta* **2006**, *69*, 334–342. [CrossRef]

26. Parashar, N.; Hait, S. Abundance, characterization, and removal of microplastics in different technology-based sewage treatment plants discharging into the middle stretch of the Ganga River, India. *Sci. Total Environ.* **2023**, *905*, 167099. [CrossRef]
27. Bayo, J.; Olmos, S.; López-Castellanos, J. Microplastics in an urban wastewater treatment plant: The influence of physicochemical parameters and environmental factors. *Chemosphere* **2020**, *238*, 124593. [CrossRef]
28. Long, Z.; Pan, Z.; Wang, W.; Ren, J.; Yu, X.; Lin, L.; Lin, H.; Chen, H.; Jin, X. Microplastic abundance, characteristics, and removal in wastewater treatment plants in a coastal city of China. *Water Res.* **2019**, *155*, 255–265. [CrossRef]
29. Bayo, J.; López-Castellanos, J. Principal factor and hierarchical cluster analyses for the performance assessment of an urban wastewater treatment plant in the Southeast of Spain. *Chemosphere* **2016**, *155*, 152–162. [CrossRef]
30. Carr, S.A.; Liu, J.; Tesoro, A.G. Transport and fate of microplastic particles in wastewater treatment plants. *Water Res.* **2016**, *91*, 174–182. [CrossRef]
31. Gies, E.A.; LeNoble, J.L.; Noël, M.; Etemadifar, A.; Bishay, F.; Hall, E.R.; Ross, P.S. Retention of microplastics in a major secondary wastewater treatment plant in Vancouver, Canada. *Mar. Pollut. Bull.* **2018**, *133*, 553–561. [CrossRef]
32. Yang, T.; Gao, M.; Nowack, B. Formation of microplastic fibers and fibrils during abrasion of a representative set of 12 polyester textiles. *Sci. Total Environ.* **2023**, *862*, 160758. [CrossRef]
33. Browne, M.A.; Crump, P.; Niven, S.J.; Teuten, E.; Tonkin, A.; Galloway, T.; Thompson, R. Accumulation of Microplastic on Shorelines Woldwide: Sources and Sinks. *Environ. Sci. Technol.* **2011**, *45*, 9175–9179. [CrossRef]
34. Yin, L.; Wen, X.; Du, C.; Jiang, J.; Wu, L.; Zhang, Y.; Hu, Z.; Hu, S.; Feng, Z.; Zhou, Z.; et al. Comparison of the abundance of microplastics between rural and urban areas: A case study from East Dongting Lake. *Chemosphere* **2020**, *244*, 125486. [CrossRef]
35. Lin, W.-H.; Ou, J.-H.; Yu, Y.-L.; Liu, P.-F.; Surampalli, R.Y.; Kao, C.-M. Regulatory Framework of Microconstituents. In *Microconstituents in the Environment*; Wiley: Hoboken, NJ, USA, 2023; pp. 513–523.
36. Tavelli, R.; Callens, M.; Grootaert, C.; Abdallah, M.F.; Rajkovic, A. Foodborne pathogens in the plastisphere: Can microplastics in the food chain threaten microbial food safety? *Trends Food Sci. Technol.* **2022**, *129*, 1–10. [CrossRef]
37. Wang, W.; Ndungu, A.W.; Li, Z.; Wang, J. Microplastics pollution in inland freshwaters of China: A case study in urban surface waters of Wuhan, China. *Sci. Total Environ.* **2017**, *575*, 1369–1374. [CrossRef]
38. Ragi, K.B.; Ekka, B.; Mezule, L. Zero pollution protocol for the recovery of cellulose from municipal sewage sludge. *Bioresour. Technol. Rep.* **2022**, *20*, 101222. [CrossRef]
39. Harley-Nyang, D.; Memon, F.A.; Jones, N.; Galloway, T. Investigation and analysis of microplastics in sewage sludge and biosolids: A case study from one wastewater treatment works in the UK. *Sci. Total Environ.* **2022**, *823*, 153735. [CrossRef]
40. Long, Y.; Zhou, Z.; Yin, L.; Wen, X.; Xiao, R.; Du, L.; Zhu, L.; Liu, R.; Xu, Q.; Li, H.; et al. Microplastics removal and characteristics of constructed wetlands WWTPs in rural area of Changsha, China: A different situation from urban WWTPs. *Sci. Total Environ.* **2022**, *811*, 152352. [CrossRef]
41. Xu, Z.; Zhai, X.; Bai, X. Amplifiers of environmental risk of microplastics in sewage sludge: Thermal drying treatment. *Sci. Total Environ.* **2023**, *905*, 167029. [CrossRef]
42. San José, R.; Pérez, J.L.; Callén, M.S.; López, J.M.; Mastral, A. BaP (PAH) air quality modelling exercise over Zaragoza (Spain) using an adapted version of WRF-CMAQ model. *Environ. Pollut.* **2013**, *183*, 151–158. [CrossRef] [PubMed]
43. Tong, H.; Jiang, Q.; Hu, X.; Zhong, X. Occurrence and identification of microplastics in tap water from China. *Chemosphere* **2020**, *252*, 126493. [CrossRef] [PubMed]
44. Halder, A.K.; Moura, A.S.; Cordeiro, M.N.D.S. Predicting the ecotoxicity of endocrine disruptive chemicals: Multitasking in silico approaches towards global models. *Sci. Total Environ.* **2023**, *889*, 164337. [CrossRef]
45. Brachi, P.; Di Fraia, S.; Massarotti, N.; Vanoli, L. Combined heat and power production based on sewage sludge gasification: An energy-efficient solution for wastewater treatment plants. *Energy Convers. Manag. X* **2022**, *13*, 100171. [CrossRef]
46. Yu, Q.; Yang, X.; Zhao, F.; Hu, X.; Guan, L.; Ren, H.; Geng, J. Spatiotemporal variation and removal of selected endocrine-disrupting chemicals in wastewater treatment plants across China: Treatment process comparison. *Sci. Total Environ.* **2022**, *835*, 155374. [CrossRef] [PubMed]
47. Sun, J.; Wang, J.; Zhang, R.; Wei, D.; Long, Q.; Huang, Y.; Xie, X.; Li, A. Comparison of different advanced treatment processes in removing endocrine disruption effects from municipal wastewater secondary effluent. *Chemosphere* **2017**, *168*, 1–9. [CrossRef]
48. Ben, W.; Zhu, B.; Yuan, X.; Zhang, Y.; Yang, M.; Qiang, Z. Transformation and fate of natural estrogens and their conjugates in wastewater treatment plants: Influence of operational parameters and removal pathways. *Water Res.* **2017**, *124*, 244–250. [CrossRef] [PubMed]
49. Mo, Q.; Yang, X.; Wang, J.; Xu, H.; Li, W.; Fan, Q.; Gao, S.; Yang, W.; Gao, C.; Liao, D.; et al. Adsorption mechanism of two pesticides on polyethylene and polypropylene microplastics: DFT calculations and particle size effects. *Environ. Pollut.* **2021**, *291*, 118120. [CrossRef] [PubMed]
50. Guart, A.; Bono-Blay, F.; Borrell, A.; Lacorte, S. Migration of plasticizersphthalates, bisphenol A and alkylphenols from plastic containers and evaluation of risk. *Food Addit. Contam. Part A* **2011**, *28*, 676–685. [CrossRef] [PubMed]
51. Edo, C.; González-Pleiter, M.; Leganés, F.; Fernández-Piñas, F.; Rosal, R. Fate of microplastics in wastewater treatment plants and their environmental dispersion with effluent and sludge. *Environ. Pollut.* **2020**, *259*, 113837. [CrossRef] [PubMed]
52. Yang, Z.; Li, S.; Ma, S.; Liu, P.; Peng, D.; Ouyang, Z.; Guo, X. Characteristics and removal efficiency of microplastics in sewage treatment plant of Xi'an City, northwest China. *Sci. Total Environ.* **2021**, *771*, 145377. [CrossRef]

53. Čelić, M.; Škrbić, B.D.; Insa, S.; Živančev, J.; Gros, M.; Petrović, M. Occurrence and assessment of environmental risks of endocrine disrupting compounds in drinking, surface and wastewaters in Serbia. *Environ. Pollut.* **2020**, *262*, 114344. [CrossRef]
54. dos Santos, D.M.; Buruaem, L.; Gonçalves, R.M.; Williams, M.; Abessa, D.M.S.; Kookana, R.; de Marchi, M.R.R. Multiresidue determination and predicted risk assessment of contaminants of emerging concern in marine sediments from the vicinities of submarine sewage outfalls. *Mar. Pollut. Bull.* **2018**, *129*, 299–307. [CrossRef]
55. Yazdan, M.M.S.; Kumar, R.; Leung, S.W. The Environmental and Health Impacts of Steroids and Hormones in Wastewater Effluent, as Well as Existing Removal Technologies: A Review. *Ecologies* **2022**, *3*, 206–224. [CrossRef]
56. Le-Minh, N.; Khan, S.J.; Drewes, J.E.; Stuetz, R.M. Fate of antibiotics during municipal water recycling treatment processes. *Water Res.* **2010**, *44*, 4295–4323. [CrossRef]
57. Almazrouei, B.; Islayem, D.; Alskafi, F.; Catacutan, M.K.; Amna, R.; Nasrat, S.; Sizirici, B.; Yildiz, I. Steroid hormones in wastewater: Sources, treatments, environmental risks, and regulations. *Emerg. Contam.* **2023**, *9*, 100210. [CrossRef]
58. Fernandes, J.P.; Almeida, C.M.R.; Pereira, A.C.; Ribeiro, I.L.; Reis, I.; Carvalho, P.; Basto, M.C.P.; Mucha, A.P. Microbial community dynamics associated with veterinary antibiotics removal in constructed wetlands microcosms. *Bioresour. Technol.* **2015**, *182*, 26–33. [CrossRef]
59. Song, H.-L.; Zhang, S.; Guo, J.; Yang, Y.-L.; Zhang, L.-M.; Li, H.; Yang, X.-L.; Liu, X. Vertical up-flow constructed wetlands exhibited efficient antibiotic removal but induced antibiotic resistance genes in effluent. *Chemosphere* **2018**, *203*, 434–441. [CrossRef]
60. Ilyas, H.; van Hullebusch, E.D. A review on the occurrence, fate and removal of steroidal hormones during treatment with different types of constructed wetlands. *J. Environ. Chem. Eng.* **2020**, *8*, 103793. [CrossRef]
61. Lee, C.-C.; Jiang, L.-Y.; Kuo, Y.-L.; Chen, C.-Y.; Hsieh, C.-Y.; Hung, C.-F.; Tien, C.-J. Characteristics of nonylphenol and bisphenol A accumulation by fish and implications for ecological and human health. *Sci. Total Environ.* **2015**, *502*, 417–425. [CrossRef]
62. Wright-Walters, M.; Volz, C.; Talbott, E.; Davis, D. An updated weight of evidence approach to the aquatic hazard assessment of Bisphenol A and the derivation a new predicted no effect concentration (Pnec) using a non-parametric methodology. *Sci. Total Environ.* **2011**, *409*, 676–685. [CrossRef]
63. iang, R.; Liu, J.; Huang, B.; Wang, X.; Luan, T.; Yuan, K. Assessment of the potential ecological risk of residual endocrine-disrupting chemicals from wastewater treatment plants. *Sci. Total Environ.* **2020**, *714*, 136689. [CrossRef]
64. Lu, S.; Lin, C.; Lei, K.; Xin, M.; Wang, B.; Ouyang, W.; Liu, X.; He, M. Endocrine-disrupting chemicals in a typical urbanized bay of Yellow Sea, China: Distribution, risk assessment, and identification of priority pollutants. *Environ. Pollut.* **2021**, *287*, 117588. [CrossRef]
65. Czarny, K.; Szczukocki, D.; Krawczyk, B.; Skrzypek, S.; Zieliński, M.; Gadzała-Kopciuch, R. Toxic effects of single animal hormones and their mixtures on the growth of Chlorella vulgaris and Scenedesmus armatus. *Chemosphere* **2019**, *224*, 93–102. [CrossRef]

toxics

MDPI

Article

White Stork Pellets: Non-Invasive Solution to Monitor Anthropogenic Particle Pollution

Dora Bjedov [1,2], Alma Mikuška [2,*], Vlatka Gvozdić [3], Petar Glavaš [2], Dora Gradečak [2] and Mirta Sudarić Bogojević [2]

1 Croatian Institute for Biodiversity, BIOTA Ltd., 10000 Zagreb, Croatia; dora.bjedov@gmail.com
2 Department of Biology, Josip Juraj Strossmayer University of Osijek, 31000 Osijek, Croatia; msudaric@biologija.unios.hr (M.S.B.)
3 Department of Chemistry, Josip Juraj Strossmayer University of Osijek, 31000 Osijek, Croatia
* Correspondence: amikuska@biologija.unios.hr; Tel.: +385-31-399-919

Abstract: The present study applied a non-invasive method to analyse anthropogenic particles and prey items in white stork (*Ciconia ciconia*) pellets. Pellets (n = 20) were obtained from white stork nests during the 2020 breeding season from two sites in Croatia. In total, 7869 anthropogenic particles were isolated. The majority of particles were fragments, while previous studies on other birds often reported fibres. An ATR–FTIR polymer analysis detected glass and construction and building materials, as well as several compounds associated with plastic masses. Polymer investigation revealed the presence of dotriacontane and octacosane, which are by-products of polyethylene (PE) degradation and transformation. Additionally, the detection of vinylidene chloride (VDC) highlights the historical contribution of polyvinylidene chloride (PVDC) to plastic pollution. Significant variation in particle quantity and size between the sampling sites was detected, with larger particles found at sites associated with the metal mechanical engineering industry and agriculture. Prey assessment revealed chitin remains of large insects such as Orthoptera and Coleoptera. This research confirms the potential of pellet analysis as a valuable tool for assessing the presence of anthropogenic particles in the environment. However, further research is needed to fully understand the extent of particle ingestion, particle sources and potential impact.

Keywords: regurgitated pellets; anthropogenic particles; pollution monitoring; dietary assessment

1. Introduction

Emerging pollutants comprise a wide category of dangerous substances, such as nanomaterials, nanoplastics, microplastics, soot and wear from roads and tyres. These anthropogenic particles are produced by human activities, resulting in their broad spatial range [1]. They are manufactured in millions of metric tonnes per year and can be released into the environment, potentially causing adverse effects on biota, the environment and public health [2]. Awareness and interest in their potentially harmful consequences have increased, especially for those at the micro- and nanoscale, e.g., organic and inorganic anthropogenic fragments [3,4]. Once in the environment, anthropogenic particles degrade into smaller particles via biotic and abiotic mechanisms, e.g., biodegradation, photodegradation, oxidation and/or abrasion [5].

Research regarding anthropogenic particle pollution has been primarily focused on the aquatic system, mainly regarding the transfer of anthropogenic particles through food webs and their effects on apex predators [6,7]. Anthropogenic particle ingestion has been previously investigated in aquatic systems via aquatic bird species, both marine (e.g., Cassin's auklet, *Ptychoramphus aleuticus* [8] and little auks, *Alle alle* [9]) and freshwater (e.g., Clapper rails, *Rallus crepitans* and Seaside sparrows, *Ammospiza maritima* [10]). However, recent studies have shown that anthropogenic particle pollution is a current ubiquitous issue [11,12]; therefore, advances have been made by analysing plastic particles in terrestrial birds. Several studies have assessed the environmental burden of anthropogenic particles in terrestrial ecosystems via white stork carcasses, focusing on general plastic ingestion [13], rubber band ingestion [14] and ingestion of plastic objects due to feeding at urban refuse dumps [15]. An additional aspect of monitoring could be accomplished by examining the quantity of anthropogenic materials utilised in the nest construction, as they can exhibit a correlation with the degree of urbanisation [16,17]. The incorporation of anthropogenic materials into nests could be affected by mating behaviour as well. Bowerbirds (Ptilonorhynchidae) construct bowers to allure potential mates [18]. The decoration of bowers plays a pivotal role in female mate selection, with bowerbirds embellishing their bowers with a variety of items, including flowers, plants and human debris such as bottle tops and straws [18]. Males with more elaborately decorated bowers are deemed more attractive and enjoy enhanced reproductive success, potentially leading to an increase in the prevalence of anthropogenic materials within bowers [18]. On the other hand, an aspect of the negative effects of anthropogenic particles was investigated, namely, the occurrence of anthropogenic materials in white stork nests, which are often associated with better breeding success. However, on the other hand, a higher risk of nestling mortality is possible due to ingestion and/or entanglement of particles [19]. Apart from lethal effects, as previously described, anthropogenic particles can cause sublethal effects, reflected in an increase in oxidative stress, overall redox imbalance and cholinesterase activity [20]. Monitoring of anthropogenic particles and their possible effects as well as integrated biomarker assessment have been used in Japanese quail, *Coturnix japonica* [20], common blackbird, *Turdus merula*, song thrush, *Turdus philomelos* [21] and tree swallow, *Tachycineta bicolor* [22], indicating the use of the aforementioned species as bioindicators of anthropogenic particle pollution in terrestrial ecosystems.

Monitoring strategies for anthropogenic particles as alternatives to bird remains include their undigested prey residues—regurgitated pellets. Pellet analysis provides information regarding prey composition as well as the occurrence of anthropogenic particles. A species that regurgitates pellets and is representative of the terrestrial ecosystem is the white stork, *Ciconia ciconia*. The species is distributed in continental Croatia [23], with opportunistic dietary habits, feeding predominantly on earthworms, grasshoppers, fish, frogs and small mammals [24]. Foraging near landfills has also been recorded [25–27]. White storks are diurnal predators, with habitat preferences in open lands, e.g., agricultural areas, wet grassland and arable lands [28]. Breeding white storks are conservative in their habitat selection, with significantly smaller home ranges, when compared to non-breeding white storks [29]. Moreover, white storks have low reproductive dispersal and usually return to the same nest as in previous years [30]. Therefore, the content of anthropogenic particles in the pellets could reflect the local environmental burden and trophic transfer.

The present research considered the white stork pellets by reporting qualitative and quantitative analysis of anthropogenic particles and fibres (plastics, textiles, construction and demolition waste and glass) in pellets from white storks. Although white stork pellets have been used for investigation to quantify their exposure to indigestible litter of anthropogenic sources and diet assessment [31], the novel aspect of this research is reflected in polymer analysis of the isolated anthropogenic particles. Therefore, the objectives of the research are as follows:

(I) Investigate the application of white stork pellets for anthropogenic particle monitoring. Since the white stork is an undomesticated species that is ecologically associated with urban settlements, their habits (behavioural and dietary) could potentially make them effective indicators of micro-anthropogenic particle pollution caused by anthropogenic activities;
(II) Perform polymer analysis on suspected anthropogenic and other non–biological particles;
(III) Examine if there is a spatial variation in the number of micro-anthropogenic particles isolated, as the assumed polluted sampling site is an area surrounded by a major river, industry and agricultural land, and is adjacent to the urban centre;

(IV) Investigate the prey composition of pellets to determine the prevalence of food sources and feeding habits of white storks in sampling locations.

2. Materials and Methods

2.1. Sampling Locations

Regurgitated pellets were obtained from white stork nests during the breeding season in June and July 2020. In total, 20 pellets were collected and analysed from two sampling areas (Figure 1). Each pellet represented one nest. Pellets from selected nests for sampling in Study Site 1 (n = 10) lay along the Sava River, just downstream from an urban centre (Slavonski Brod) known for its highly developed metal engineering industry. The nests are surrounded by agriculture, small villages, alluvial forests and pastures regularly flooded by Sava. Furthermore, an oil refinery is situated at Bosanski Brod, which is adjacent to the town of Slavonski Brod. Pellets sampled from nests in Study Site 2 (n = 10) were located in small villages, surrounded by large grassland pastures, meadows, arable land and woodland habitat.

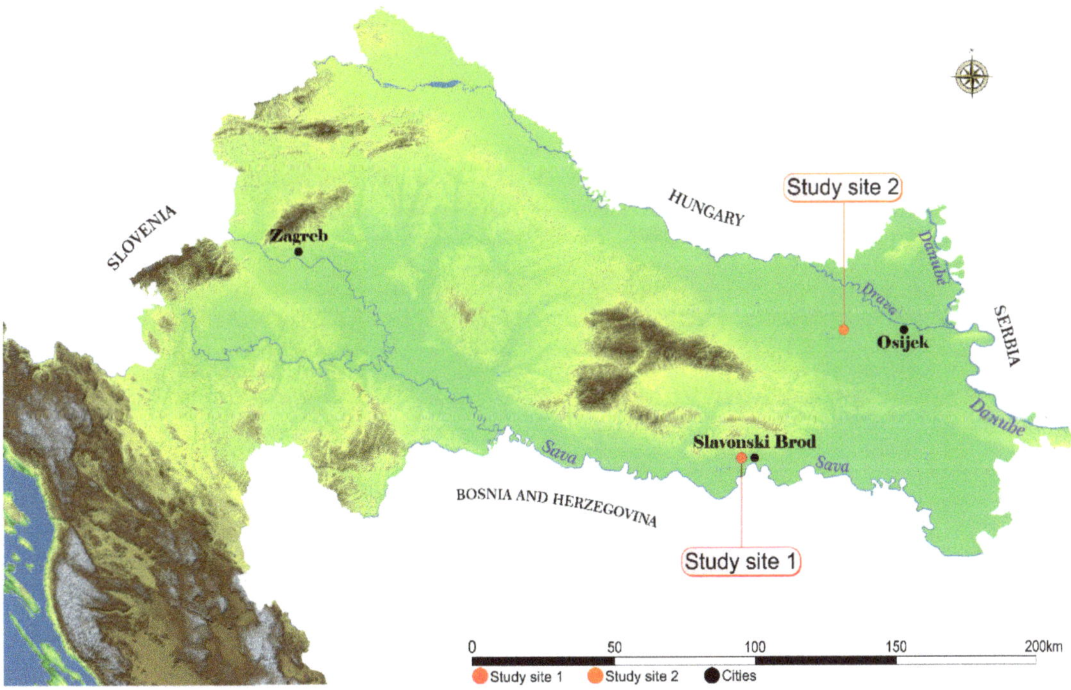

Figure 1. The geographical location of sampling sites.

2.2. Isolation and Analysis of Anthropogenic Particles

Following the field sampling, all pellets were kept at −20 °C to prevent microbial growth until analysis. Thawed pellets were weighed and dissected. Potential anthropogenic particles were visually detected with a high-quality stereomicroscope Leica MZ6 and categorised by size into microanthropogenic (< 0.50 mm) and macroanthropogenic particles (>0.50 mm). Another category of the particles was shape (e.g., fragment, filament). Suspected isolated particles were subsequently corroborated with the hot needle method. The hot needle method has been used as the visual verification prior to advanced polymer identification. To expand, a histological needle was heated on a glass alcohol burner and put on the suspected particle of anthropogenic origin. A positive response was observed if the particle melted or curled, rather than charred [32,33]. The detection of microplastic particles smaller than 0.5 mm was performed based on shape and colour with an optical microscope (Leica MZ6). The isolated micro– and macroanthropogenic particles were transferred to glass vials with metal tweezers and stored until analysis.

2.3. Spectroscopic Analysis

Polymer analysis of isolated particles was performed with attenuated total reflection Fourier transform infrared spectroscopy (ATR–FTIR). In total, 642 particles were selected for analysis based on the hot needle test, size, shape and colour. Anthropogenic particles were analysed with ATR–FTIR in a wavenumber range of 4000–450 cm^{-1}. Each sample was measured in six technical replicates. The obtained spectrum for each sample was recorded as % transmittance (T) using a Perkin-Elmer Spectrum Two with Universal ATR, controlled by the software Spectrum 10.5.2.636.

2.4. Prey Remains Isolation and Determination

In parallel with anthropogenic particles, prey remains were isolated with dry method pellet analysis, according to Horváth et al. [34]. The identification of prey was based on the morphological characteristics of the remains. Prey items were identified at the lowest possible taxonomic level. Chitinous pieces of insects were identified according to Chinery [35] and by comparison with entomological collections of species commonly present in the studied areas.

2.5. Quality Control

Quality control precautions were implemented during the isolation and polymer analysis of anthropogenic particles. Plastic materials were intentionally avoided throughout the process of pellet collection, sample isolation and sample analysis. Instead, preference was given to the use of glass vials and Petri dishes, as well as aluminium and stainless-steel utensils, for all equipment. Additionally, lab coats and nitrile gloves were worn, samples were covered with aluminium foil when not being used or processed and procedural blanks were used. Particles were isolated in a laminar flow cabinet equipped with vertical HEPA filters (MINIFLO Type 90, Milan, Italy). The laboratory workspace as well as tweezers, needles, glass vials and Petri dishes were meticulously cleaned with 70% ethyl alcohol.

2.6. Statistical Analysis

Statistical tests were performed using R version 4.2.2 and Statistica version 14.0.0.15. To identify the patterns and/or trends in the data that may indicate variations in polymer composition with regard to sampling sites, principal component analysis (PCA) was performed. To compare the number of isolated anthropogenic particles with regard to sampling sites, the number of particles per mass of the pellet ($n_{particle}\ g_{pellet}{}^{-1}$) was used. To test the normality of the data distribution, the Shapiro–Wilks test was applied. Data were not normally distributed; therefore, the non-parametric, unpaired, two-tailed Mann–Whitney U test was applied by comparing the ranks. The level of statistical significance (p-value) was 0.05 throughout the study.

3. Results

3.1. Isolated Anthropogenic Particles

Anthropogenic particles were detected in all analysed pellets. Particles such as microplastic fragments, filaments, building materials and glass were isolated and morphological characteristics were determined. More than 90% of anthropogenic particles were clear fragments, followed by filaments (Figure 2). Microanthropogenic particles were detected in all pellets, while macroanthropogenic particles were reported in 60% of analysed pellets.

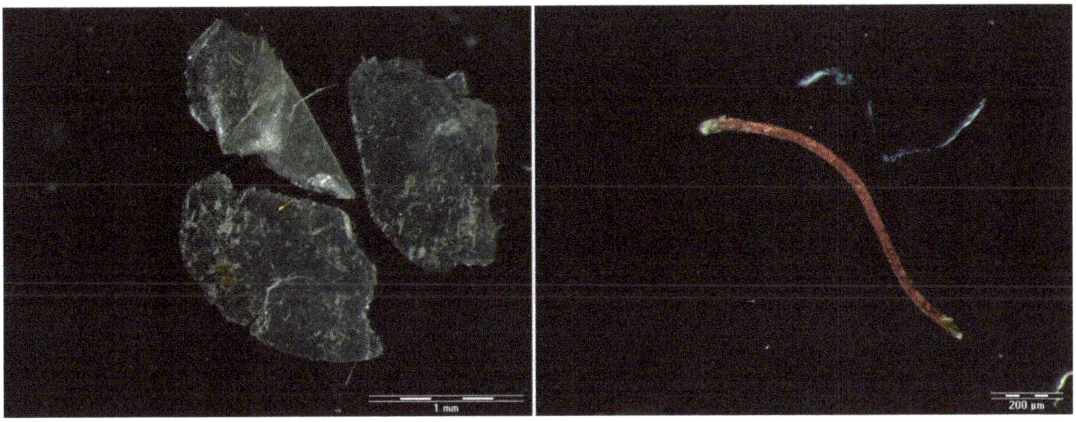

Figure 2. The most common anthropogenic particles found in white stork (*C. ciconia*) pellets were clear fragments (**left**; compound dotriacontane) and coloured filaments (**right**; compound paraffin oil).

3.2. ATR–FTIR Results of Analysed Particles

We detected substances associated with plastic masses, which are shown in Table 1. Out of 7869 isolated particles, we detected polymers associated with plastic masses in 519 particles, namely, in Study Site 1, 321 (4.23%) particles and, in Study Site 2, 198 (49.38%) were associated with plastic masses. PCA analysis was performed on 499 spectra. The results showed similarities among samples collected from different locations. The results of PCA showed that two principal components account for 80% of the total variance in the data (Figure 3).

Table 1. Results of polymers detected with ATR–FTIR. For each polymer, a use was described as well as whether it is associated with plastic masses.

Polymer	Uses	Associated with Plastic Masses
(3-aminopropyl)triethoxysilane	thermoplastic polymer	yes
1,2-octadecanediol	personal care products	no
1,3,5-trimethylcyclohexane	by-product of PE	yes
1-chlorohexadecane	additive used in plastic production	yes
3-(2-imidazolin-1-YL)propyltriethoxysilane	resin and plastic production	yes
3-methylheptane	product of PS degradation	yes
Butyl stearate	additive used in plastic production	yes
Dioctyl sebacate	additive used in plastic production	yes
Dotriacontane	by-product of PE	yes
Enzacryl polyacetal	thermoplastic polymer	yes
Ethyl palmitate	product of PU degradation	yes
Hexacosanol	plastic production	yes
Hexatriacontane	petroleum product	no
L(-)-glyceraldehyde unnatural forms	naturally occurring	no
Methyl linoleate	PVC plasticiser	yes
Octacosane	by-product of PE	yes
Octadecylamine	product of PU degradation	yes
Paraffin oil	plastic production	yes
Polystyrene	plastic polymer	yes
Tetradodecylammonium bromide	surfactant and catalyst	no
Toluene-4-sulfonic acid	surfactant and catalyst	no
Vinylidene chloride	plastic production	yes

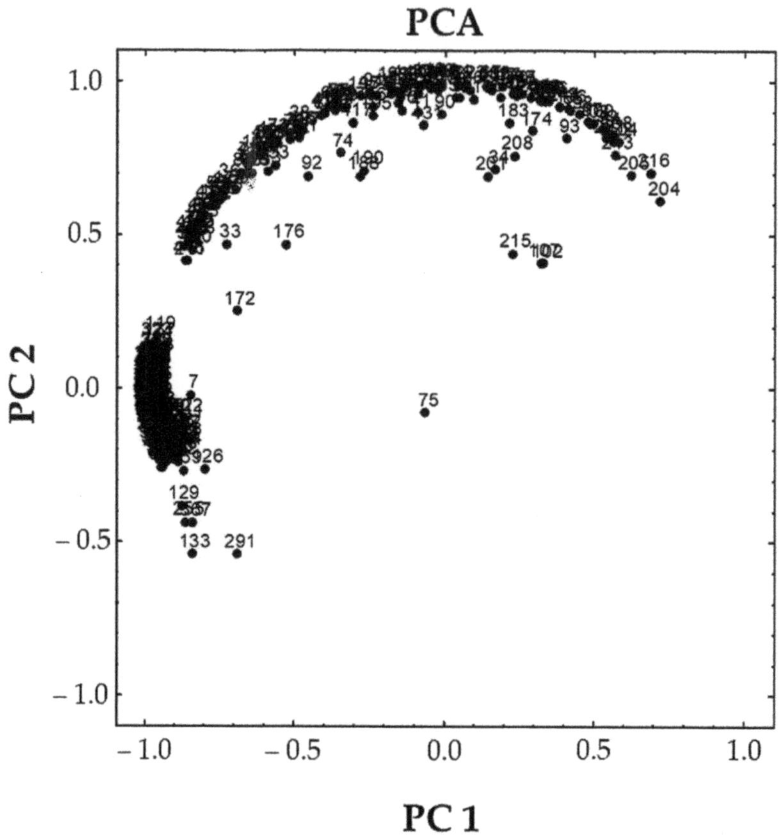

Figure 3. The results of principal component analysis (PCA): one complex cluster of all analysed spectra.

3.3. Spatial Variability

In total, 7869 micro- and macroanthropogenic particles were isolated from the pellets. An overview of the results is shown in Table 2. In particular, there were 7869 particles in Study Site 1 and 401 in Study Site 2, confirming the significant spatial variation in particle quantity ($p < 0.0001$). Regarding particle size, significantly larger particles were detected in Study Site 1 compared to Study Site 2 (Table 2).

Table 2. Number, mass and diameter of isolated anthropogenic particles from white stork (*C. ciconia*) pellets sampled during breeding season 2020 in continental Croatia.

	$n_{particle}$	Mass (g)	$n_{particle}$ g_{pellet}^{-1}	Min	Max	Mean ± SD
Study Site 1 (n_{pellet} = 10)	284	13.23	21.47	<0.50	20.00	2.54 ± 1.68
	239	6.25	38.22	1.00	40.00	2.27 ± 3.10
	33	11.26	2.93	1.00	10.00	3.12 ± 1.68
	105	12.28	8.55	1.00	10.00	2.10 ± 1.21
	86	12.28	7.00	<0.50	10.00	2.37 ± 1.30
	660	9.00	73.37	1.00	22.00	2.39 ± 0.93
	1411	27.93	50.51	<0.50	13.00	2.33 ± 0.93
	796	7.20	110.49	<0.50	7.00	2.02 ± 0.93
	1996	22.08	90.39	<0.50	20.00	1.80 ± 1.29
	1858	13.51	137.51	<0.50	12.00	2.32 ± 1.26

Table 2. *Cont.*

	$n_{particle}$	Mass (g)	$n_{particle}$ g_{pellet}^{-1}	Min	Max	Mean ± SD
Study Site 2 (n_{pellet} = 10)	51	9.63	5.30	<0.50	5.00	1.37 ± 0.91
	27	10.73	2.52	<0.50	3.00	1.28 ± 0.71
	35	17.16	2.04	<0.50	1.00	0.73 ± 0.24
	33	7.53	4.38	<0.50	2.25	0.93 ± 0.46
	125	11.51	10.86	<0.50	4.25	1.32 ± 0.83
	12	8.24	1.46	<0.50	1.20	0.76 ± 0.22
	12	4.80	2.50	<0.50	2.50	1.25 ± 0.58
	4	11.87	0.34	<0.50	1.20	0.85 ± 0.31
	9	8.95	1.01	<0.50	35.00	5.83 ± 11.23
	93	7.23	12.87	<0.50	9.00	1.84 ± 1.46

$n_{particle}$—number of isolated anthropogenic particles; Mass—the mass of the whole dry pellet; $n_{particle}$ g_{pellet}^{-1}—number of isolated anthropogenic particles per gram of the pellet; Min—minimum diameter of particles in the pellet; Max—maximum diameter of particles in the pellet.

3.4. Dietary Assessment

Pellet analyses showed that white storks from study locations fed on insects (Insecta), spiders (Arachnida), snails (Gastropoda), earthworms (Clitellata) and mammals (Mammalia). In all analysed pellets, remains of mammals' hair and earthworms' chaetae were found (Table 3), along with different blades of grass and other plants. Among insects, the most abundant prey remains belonged to beetles (Coleoptera), grasshoppers, locusts and crickets (Orthroptera). Differences between prey remains from the two study sites are presented in Table 3.

Table 3. Taxonomic groups of prey items determined in the pellets of white stork (*C. ciconia*) and their occurrence in Study Site 1 and Study Site 2.

Class	Order	Family	Species	Study Site 1	Study Site 2
Mammalia	Rodentia			x	x
Arachnida	Araneae			x	
Clitellata	Opisthopora	Lumbricidae		x	x
Mollusca	Gastropoda		*Gastropoda terrestria* sp.	x	
Insecta	Diptera				x
	Hymenoptera	Formicidae		x	
	Orthoptera	Gryllidae		x	x
		Tettigoniidae			x
		Acrididae			x
		Gryllotalpidae	*Gryllotalpa gryllotalpa*	x	x
	Coleoptera	Chrysomelidae		x	
		Silphidae/		x	x
		Lucanidae	*Dorcus parallelipipedus*	x	x
		Cerambycidae			x
		Tenebrionidae	*Blaps mortisaga*	x	
		Scarabaeidae	*Melolontha* sp.		x
			Melolontha melolontha	x	
			Oryctes nasicornis	x	
			Cetonia aurata	x	
		Carabidae	*Carabus* sp.	x	x
			Abax sp.	x	x
			Calosoma sp.		x
			Harpalus sp.		x
			Abax sp.	x	x

Table 3. *Cont.*

Class	Order	Family	Species	Study Site 1	Study Site 2
			Carabus ullrichi Germar	x	x
			Carabus granulatus	x	
			Carabus violaceus	x	
			Carabus coriaceus	x	
			Carabus intricatus	x	
			Calosoma auropunctatum	x	

4. Discussion

The present study implemented the method of collecting and analysing white stork pellets for the purpose of anthropogenic particle monitoring. Among species that regurgitate pellets, anthropogenic and plastic particles have been detected in white stork, kingfisher, *Alcedo atthis* and barn owl (*Tyto alba*) [31,36,37]. That being said, our results correspond with the study by Mikula et al. [26], as we detected anthropogenic particles in all analysed pellets as well. Anthropogenic particles were also detected in earlier examinations of white stork pellets collected in Bulgaria during the non-breeding season, albeit at far lower frequencies in pellets for glass (2.7%) and plastic (4.1%) [38]. Nessi et al. [37] analysed microplastics in the pellets of a nocturnal bird of prey, barn owl. The authors associated the microplastic from the pellets with prey due to degradation of habitat, i.e., agricultural lands [37]. In research on kingfisher, a piscivore top predator in river ecosystems, the authors suggested that the ingestion was more likely derived from their food rather than from abiotic elements such as sediment and water [36]. Research on waterbirds suggests the ingestion of microplastics likely originates from sediment particles and water rather than from their food, although this has yet to be conclusively proven [39]. Regarding shape, most detected particles from other studies were fibres [36,37,40], while, in the present study, most detected particles were fragments. Anthropogenic micro–fragments can be derived from the breakdown, fragmentation or degradation of larger anthropogenic particles [41]. Although results from the present study are difficult to compare to other studies due to different avian foraging strategies, pellet regurgitation, habitat, research methodology and pollutant accumulation, continuous detection of anthropogenic particles in pellets, digestive tract and faeces indicates environmental pollution, warranting design of mitigation measures. When interpreting results, several sources of anthropogenic particles in pellets should be taken into account. For example, particles can be ingested primarily by accident together with smaller food items such as insects or secondarily if the anthropogenic particles are digested by their prey. An additional source of particles in sampled pellets could be atmospheric deposition [42].

PCA results did not show any significant clustering of the polymer compounds based on the sampling site variable. Anthropogenic pollution appears similar in a polymer sense but differs in quantity, as seen by the number of isolated particles per site. According to Moore [43], the polymers found in microplastic pollutants can undergo degradation and possible chemical changes due to exposure to the environment. Furthermore, Lundquist et al. [44] suggest that microplastic pollutants consist of various inorganic fillers, plasticisers and UV stabilisers, which may also undergo alterations caused by environmental conditions. The ATR–FTIR spectra of a microplastic particle will reflect all the chemical changes it has experienced, including the presence of non–polymer compounds from the pollutant. However, it is crucial to consider the presence of typical additives and co–polymers that might also be present when interpreting the results.

The most common chemical compounds when analysing isolated microparticles were dotriacontane and octacosane. According to Abraham et al. [45], dotriacontane is a by-product of plastic polymer polyethylene (PE) degradation by fungi, *Aspergillus nomius*. Octacosane and 1,3,5-trimethylcyclohexane are by-products of low-density PE transformation under high temperatures [46,47]. Since PE is a polymer that is primarily used for packaging, e.g., plastic bags, films and containers, this represents the first association with microplastic particles in white stork pellets. Several other compounds associated with plastic masses have been detected. Compound 3-(2-imidazolin-1-YL) propyltriethoxysilane is used in resin and plastic production [48] and methyl linoleate is a plasticiser used for polyvinyl chloride (PVC) [49]. Enzacryl polyacetal is a synthetic polymer, a thermoplastic used in engineering. Previously, it has been characterised only in aquatic ecosystems, namely, two fish: Spotted Tail goby, *Synechogobius ommaturus*, and Seabass, *Lateolabrax japonicus* [6]. Another thermoplastic compound detected was (3-aminopropyl)triethoxysilane. Additionally, we detected vinylidene chloride (VDC, 1,1-dichloroethylene), a compound used in the production of the polymer polyvinylidene chloride (PVDC). PVDC is well known for its barrier properties and is used extensively as a coating for various packaging materials, especially in the food industry. It is often used in combination with other polymers to create materials with enhanced barrier properties against moisture, oxygen and other gases [50]. While PVDC itself is not as commonly used today due to environmental and health concerns related to the release of vinyl chloride monomer during production and incineration, it has historically been a significant contributor to plastic pollution [51]. Polymer analysis revealed paraffin oil on the analysed particles. Paraffin oil has many uses in the plastic industry and is associated with agriculture, e.g., petroleum-based insecticides and as a part of diesel fuel for tractor engines [52]. White storks are frequently associated with foraging on arable lands; therefore, it is no surprise the residues of agricultural and farming equipment have been detected. Chemicals obtained by bacterial degradation of chlorinated paraffins were observed. Dioctyl sebacate and 1-chlorohexadecane are examples of additives used in plastics to modify certain properties or facilitate the manufacturing process, namely, 1-chlorohexadecane was detected and its main purpose is industrial. It is frequently added in plasticisers and flame retardants [53]. Apart from chemicals associated with plastic degradation, compounds (hexacosanol) used in plastic production as molecular lubricants for plastic polymers were detected [54]. In particular, butyl stearate is used as a functional additive, acting as a lubricant in the plastic polymer polystyrene (PS). Volatile organic compounds (VOCs; e.g., 3-methylheptane) have been detected. VOCs are usually released in the environment by photodegradation of various plastic polymers, such as PS [55]. As previously mentioned, visual inspection of macroanthropogenic particles showed construction and building materials in the pellets. This was additionally confirmed by ATR–FTIR analysis of particles that contained octadecylamine. Octadecylamine is a compound associated with the improvement of the hydrophobic properties of polyurethane (PU) foam for the purpose of oil spill clean-up [56]. Ethyl palmitate was detected as well. The compound is a degradation product of PU [57]. Hydrocarbons were detected in the pellets as well. Hexatriacontane indicates the presence of these persistent organic pollutants (POP) derived from petroleum and contributes to environmental pollution and adverse effects on biota [58,59]. Potential sources of hexatriacontane are motorised activities and the petrochemical industry [58,60].

Anthropogenic particles obtained from regurgitated pellets from white storks' nests at Study Site 1 and Study Site 2, varied significantly in particle quantity (Table 1). Regarding particle size, significantly larger particles were detected in Study Site 1 compared to Study Site 2 (Table 1). The white stork forages on open grasslands and floodplains, habitats often transformed into agricultural and farming lands. Agricultural soils may become long–term 'sinks' and reservoirs for anthropogenic particles [61,62]. This indicates that agricultural areas are vulnerable to pollution, reflected in anthropogenic particle detection in both study areas. However, a greater number of (and larger) man-made particles were detected in pellets from Study Site 1. We assume that the city and the urban residential area actively contribute to the anthropogenic particle pollution, based on the fact that microplastic particles have been detected in soil and surface road dust in urban cities [63,64]. Since the foraging area is in proximity to the urban centre of Slavonski Brod, the wastewater treatment plant (WWTP) in Slavonski Brod can be a potential source of anthropogenic particles via the release of effluent plants [65]. Furthermore, the metallurgic industry in

Slavonski Brod and the oil refinery in Bosanski Brod could be major potential sources of pollution in the Sava River and the surrounding soil.

White storks regurgitate pellets daily or even more times per day, depending on prey abundance [66]. Foraging flights of the majority of white storks are within 1.5 km of nests [67,68], but foraging radius can be up to 5 km from nests [69]. The diversity of prey items depends on the conditions prevailing in their habitats—if the habitats are dry and there is no larger prey available, white storks will feed on insects [70]. Depending on the type of prey, white storks have different hunting strategies. They catch their prey with their long beaks, and, if it is a larger animal, they first kill it with a beak strike and then tear it apart. Insects are collected by searching through low vegetation [66]. In the dietary assessment, we found only small mammal hairs (from which it is not possible to determine species, number of specimens or their size) and no remains from fishes, amphibians or reptiles. Studies of white stork feeding habits show that the deficiency of prey remains of mammals, amphibians, reptiles or fishes in pellets does not reflect a lack of them in the feeding habitats, but rather that their remains are almost entirely digested [68,70]. We found numerous chitin remains of large insects—mandibles from Orthoptera and elytrons from Coleoptera. Our results comply with diet studies in Europe showing that insects are important prey for white storks, especially in southern parts of Europe where habitats are drier [71–73].

5. Conclusions

The present research successfully applied the pellets of an opportunistic terrestrial apex predator for anthropogenic particle monitoring. The findings suggest that pellet analysis offers a non–invasive method to assess the presence of various pollutants in the environment while reducing disturbance and minimising ethical concerns. Following a polymer analysis, we detected construction and building materials, glass and several compounds associated with plastic masses. The ATR–FTIR analysis of isolated particles revealed the presence of dotriacontane and octacosane, which are by-products of PE degradation and transformation. Additionally, the detection of VDC highlights the historical contribution of PVDC to plastic pollution. Regarding quantity, spatial variation was confirmed, as a higher number of fragments was detected from pellets in Study Site 1. It is assumed that the wastewater treatment plant in Slavonski Brod contributes to the high number of fragments. Diet assessment of the white stork revealed a lack of identifiable remains from fishes, amphibians or reptiles, suggesting efficient digestion, while chitin remains of large insects such as Orthoptera and Coleoptera were abundant. To conclude, the presence of man-made fragments in white stork pellets highlights the problem of widespread anthropogenic particles in the environment. By analysing the composition and characteristics of the particles found in the pellets, it is possible to identify specific pollutants, their origins and pollutant hotspots, making storks valuable indicator species for environmental monitoring. Analysis of pellets over time offers a valuable means to elucidate temporal variations in pollutant concentrations and trends, thereby facilitating a comprehensive understanding of pollution dynamics within the ecosystem. Such insights are instrumental in informing, formulating and refining policies and regulations that are targeted at mitigating particle pollution, ultimately contributing to environmental management and public health enhancement efforts. Additionally, the chemical compounds associated with anthropogenic and plastic debris and the analysis of anthropogenic particles (as well as microplastics) should be considered in future research to understand their effect on biota and their role in the ecosystem, if any.

Author Contributions: Conceptualisation, D.B. and A.M.; formal analysis, D.B., P.G. and D.G.; investigation, D.B., V.G., P.G., D.G. and M.S.B.; resources, A.M. and M.S.B.; writing—original draft preparation, D.B.; writing—review and editing, All; visualisation, D.B.; supervision A.M., V.G. and M.S.B.; funding acquisition, A.M. and M.S.B. All authors have read and agreed to the published version of the manuscript.

Funding: This research was financially and materially supported by the Department of Biology's project Wetland ornithofauna research: the presence of non-essential metals and anthropogenic particles in the diet of white storks and white-tailed eagles in Croatia (Grant number OZB-ZP2022-3105422).

Institutional Review Board Statement: Samples and data were collected according to Institute of Ornithology, Croatian Academy of Science, protocols, under the supervision of a certified ringer/researcher. Samples and data were collected as part of the routine White Stork ringing and monitoring scheme in Republic of Croatia. All procedures were conducted in accordance with the Croatian Nature Protection Act [Official Gazette no. 80/13, 15/18 and 14/19] and approved by the Croatian Ministry of Economy and Sustainable Development (Classification code: UP/I-612-07/20-48/130; Registry number: 517-05-1-1-20-4). No extra animal discomfort was caused during sample collection for the purpose of this study.

Data Availability Statement: All data are included in the manuscript.

Acknowledgments: The authors gratefully acknowledge the invaluable support of Vlatko Galić, PhD and Lovro Vukadinović, mag. prot. nat. et amb., from the Agricultural Institute of Osijek, who generously provided the necessary equipment for conducting the polymer analysis essential to this research (Project: Support for conservation, sustainable use and development of genetic resources in agriculture, PRR-10.02.01.00.0-06).

Conflicts of Interest: The authors declare no conflicts of interest.

References

1. Mattsson, K.; da Silva, V.H.; Deonarine, A.; Louie, S.M.; Gondikas, A. Monitoring anthropogenic particles in the environment: Recent developments and remaining challenges at the forefront of analytical methods. *Curr. Opin. Colloid. Interface Sci.* **2021**, *56*, 101513. [CrossRef]
2. Thompson, R.C.; Olsen, Y.; Mitchell, R.P.; Davis, A.; Rowland, S.J.; John, A.W.; McGonigle, D.; Russell, A.E. Lost at sea: Where is all the plastic? *Science* **2004**, *304*, 838. [CrossRef] [PubMed]
3. Koelmans, A.A.; Besseling, E.; Foekema, E.; Kooi, M.; Mintenig, S.; Ossendorp, B.C.; Redondo-Hasselerharm, P.E.; Verschoor, A.; van Wezel, A.P.; Scheffer, M. Risks of Plastic Debris: Unravelling Fact, Opinion, Perception, and Belief. *Environ. Sci. Technol.* **2017**, *51*, 11513–11519. [CrossRef] [PubMed]
4. Koelmans, A.A.; Redondo-Hasselerharm, P.E.; Nor, N.H.M.; de Ruijter, V.N.; Mintenig, S.M.; Kooi, M. Risk assessment of microplastic particles. *Nat. Rev. Mater.* **2022**, *7*, 138–152. [CrossRef]
5. Auta, H.S.; Emenike, C.U.; Jayanthi, B.; Fauziah, S.H. Growth kinetics and biodeterioration of polypropylene microplastics by *Bacillus* sp. and *Rhodococcus* sp. isolated from mangrove sediment. *Mar. Pollut. Bull.* **2018**, *127*, 15–21. [CrossRef] [PubMed]
6. Wang, F.; Wu, H.; Wu, W.; Wang, L.; Liu, J.; An, L.; Xu, Q. Microplastic characteristics in organisms of different trophic levels from Liaohe Estuary, China. *Sci. Total Environ.* **2021**, *789*, 148027. [CrossRef] [PubMed]
7. Zheng, X.; Wu, X.; Zheng, Q.; Mai, B.X.; Qiu, R. Transfer of Microplastics in Terrestrial and Aquatic Food Webs: The Impact of E-Waste Debris and Ecological Traits. *Environ. Sci. Technol.* **2022**, *57*, 1300–1308. [CrossRef] [PubMed]
8. O'Hara, P.D.; Avery-Gomm, S.; Wood, J.; Bowes, V.; Wilson, L.; Morgan, K.H.; Boyd, W.S.; Hipfner, J.M.; Desforges, J.-P.; Bertram, D.F.; et al. Seasonal variability in vulnerability for Cassin's auklets (*Ptychoramphus aleuticus*) exposed to microplastic pollution in the Canadian Pacific region. *Sci. Total Environ.* **2019**, *649*, 50–60. [CrossRef] [PubMed]
9. Amélineau, F.; Bonnet, D.; Heitz, O.; Mortreux, V.; Harding, A.; Karnovsky, N.; Walkusz, W.; Fort, J.; Grémillet, D. Microplastic pollution in the Greenland Sea: Background levels and selective contamination of planktivorous diving seabirds. *Environ. Pollut.* **2016**, *219*, 1131–1139. [CrossRef]
10. Weitzel, S.L.; Feura, J.M.; Rush, S.A.; Iglay, R.B.; Woodrey, M.S. Availability and assessment of microplastic ingestion by marsh birds in Mississippi Gulf Coast tidal marshes. *Mar. Pollut. Bull.* **2021**, *166*, 112187. [CrossRef]
11. Wong, J.K.H.; Lee, K.K.; Tang, K.H.D.; Yap, P.S. Microplastics in the freshwater and terrestrial environments: Prevalence, fates, impacts and sustainable solutions. *Sci. Total Environ.* **2020**, *719*, 137512. [CrossRef]
12. Campanale, C.; Galafassi, S.; Savino, I.; Massarelli, C.; Ancona, V.; Volta, P.; Uricchio, V.F. Microplastics pollution in the terrestrial environments: Poorly known diffuse sources and implications for plants. *Sci. Total Environ.* **2022**, *805*, 150431. [CrossRef]
13. Nicastro, K.R.; Savio, R.L.; McQuaid, C.D.; Madeira, P.; Valbusa, U.; Azevedo, F.; Casero, M.; Lourenço, C.; Zardi, G.I. Plastic ingestion in aquatic-associated bird species in southern Portugal. *Mar. Pollut. Bull.* **2018**, *126*, 413–418. [CrossRef] [PubMed]
14. Henry, P.Y.; Wey, G.; Balança, G. Rubber Band Ingestion by a Rubbish Dump Dweller, the White Stork (*Ciconia ciconia*). *Waterbirds* **2011**, *34*, 504–508. [CrossRef]
15. Peris, S. Feeding in urban refuse dumps: Ingestion of plastic objects by the White Stork (*Ciconia ciconia*). *Ardeola* **2003**, *50*, 81–84.
16. Radhamany, D.; Das, K.S.; Azeez, P.A.; Wen, L.; Sreekala, L.K. Usage of nest materials by house sparrow (*Passer domesticus*) along an urban to rural gradient in Coimbatore, India. *Trop. Life Sci. Res.* **2016**, *27*, 127. [CrossRef] [PubMed]

17. Wang, Y.; Chen, S.; Blair, R.B.; Jiang, P.; Ding, P. Nest composition adjustments by Chinese Bulbuls *Pycnonotus sinensis* in an urbanized landscape of Hangzhou (E China). *Acta Ornithol.* **2009**, *44*, 185–192. [CrossRef]
18. Borgia, G. Bower quality, number of decorations and mating success of male satin bowerbirds (*Ptilonorhynchus violaceus*): An experimental analysis. *Animal Behaviour* **1985**, *33*, 266–271. [CrossRef]
19. Jagiello, Z.A.; Dylewski, Ł.; Winiarska, D.; Zolnierowicz, K.M.; Tobolka, M. Factors determining the occurrence of anthropogenic materials in nests of the white stork *Ciconia ciconia*. *Environ. Sci. Pollut. Res. Int.* **2018**, *25*, 14726–14733. [CrossRef]
20. de Souza, S.S.; Freitas, N.; Gonçalves, S.d.O.; da Luz, T.M.; Araújo, A.P.d.C.; Rajagopal, R.; Balasubramani, G.; Rahman, M.; Malafaia, G. Toxicity induced via ingestion of naturally-aged polystyrene microplastics by a small-sized terrestrial bird and its potential role as vectors for the dispersion of these pollutants. *J. Hazard. Mater.* **2022**, *434*, 128814. [CrossRef]
21. Deoniziak, K.; Cichowska, A.; Niedźwiecki, S.; Pol, W. Thrushes (Aves: Passeriformes) as indicators of microplastic pollution in terrestrial environments. *Sci. Total Environ.* **2022**, *853*, 158621. [CrossRef]
22. Sherlock, C.; Fernie, K.J.; Munno, K.; Provencher, J.; Rochman, C. The potential of aerial insectivores for monitoring microplastics in terrestrial environments. *Sci. Total Environ.* **2022**, *807 Pt 1*, 150453. [CrossRef]
23. Kralj, J.; Barišić, S.; Tutiš, V.; Ćiković, D. *Croatian Bird Migration Atlas*; Kralj, J., Barišić, S., Tutiš, V., Ćiković., D., Eds.; Atlas selidbe ptica Hrvatske; Hrvatska akademija znanosti i umjetnosti (HAZU): Zagreb, Croatia, 2013.
24. Kosicki, J.Z.; Profus, P.; Dolata, P.T.; Tobółka, M. Food composition and energy demand of the White Stork *Ciconia ciconia* breeding population. Literature survey and preliminary results from Poland. In *The White Stork in Poland: Studies in Biology, Ecology and Conservation*; Bogucki Wydawnictwo Naukowe: Poznań, Poland, 2006; pp. 169–183.
25. Bjedov, D.; Velki, M.; Lackmann, C.; Begović, L.; Mikuška, T.; Jurinović, L.; Mikuška, A. Blood biomarkers in white stork (*Ciconia ciconia*) nestlings show different responses in several areas of Croatia. *J. Exp. Zool. A Ecol. Integr. Physiol.* **2022**, *337*, 547–558. [CrossRef] [PubMed]
26. de la Casa-Resino, I.; Hernández-Moreno, D.; Castellano, A.; Pérez-López, M.; Soler, F. Breeding near a landfill may influence blood metals (Cd, Pb, Hg, Fe, Zn) and metalloids (Se, As) in white stork (*Ciconia ciconia*) nestlings. *Ecotoxicology* **2014**, *23*, 1377–1386. [CrossRef] [PubMed]
27. López-García, A.; Sanz-Aguilar, A.; Aguirre, J.I. The trade-offs of foraging at landfills: Landfill use enhances hatching success but decrease the juvenile survival of their offspring on white storks (*Ciconia ciconia*). *Sci. Total Environ.* **2021**, *778*, 146217. [CrossRef]
28. Radović, A.; Tepić, N. Using Corine Land Cover habitat database for the analysis of breeding bird habitat: Case study of white storks (*Ciconia ciconia*) from northern Croatia. *Biologia* **2009**, *64*, 1212–1218. [CrossRef]
29. Zurell, D.; von Wehrden, H.; Rotics, S.; Kaatz, M.; Groß, H.; Schlag, L.; Schäfer, M.; Sapir, N.; Turjeman, S.; Wikelski, M.; et al. Home Range Size and Resource Use of Breeding and Non-breeding White Storks Along a Land Use Gradient. *Front. Ecol. Evol.* **2018**, *6*, 79. [CrossRef]
30. Ječmenica, B.; Kralj, J. Dispersal of the White Stork *Ciconia ciconia* in the Lonjsko polje Nature Park, Croatia. *Larus Godišnjak Zavoda Za Ornitol. Hrvat. Akad. Znan. I Umjet.* **2017**, *52*, 35–48. [CrossRef]
31. Mikula, P.; Karg, J.; Jerzak, L.; Walasz, K.; Siekiera, J.; Czyż, S.; Mikicińska, K.; Pietkiewicz, M.; Sztwiertnia, H.; Wyka, J.; et al. Diet analysis and the assessment of plastic and other indigestible anthropogenic litter in the white stork pellets. *Environ. Sci. Pollut. Res.* **2023**, *31*, 6922–6928. [CrossRef]
32. De Witte, B.; Devriese, L.; Bekaert, K.; Hoffman, S.; Vandermeersch, G.; Cooreman, K.; Robbens, J. Quality assessment of the blue mussel (*Mytilus edulis*): Comparison between commercial and wild types. *Mar. Pollut. Bull.* **2014**, *85*, 146–155. [CrossRef]
33. Beckingham, B.; Apintiloaiei, A.; Moore, C.; Brandes, J. Hot or not: Systematic review and laboratory evaluation of the hot needle test for microplastic identification. *Microplastics Nanoplastics* **2023**, *3*, 8. [CrossRef]
34. Horváth, G.; Jurčević Agić, I.; Merdić, E.; Torizs, I.; Purger, J. Monitoring sitnih sisavaca na temelju istraživanja sastava gvalica sova. In *Priručnik za istraživanje bioraznolikosti duž rijeke Drave Sveučilište u Pečuhu*; University of Pécs: Pécs, Hungary, 2007; pp. 203–217.
35. Chinery, M. *Insects of Britain and Northern Europe*, 3rd ed.; HarperCollins: London, UK, 1993.
36. Winkler, A.; Nessi, A.; Antonioli, D.; Laus, M.; Santo, N.; Parolini, M.; Tremolada, P. Occurrence of microplastics in pellets from the common kingfisher (*Alcedo atthis*) along the Ticino River, North Italy. *Environ. Sci. Pollut. Res. Int.* **2020**, *27*, 41731–41739. [CrossRef]
37. Nessi, A.; Winkler, A.; Tremolada, P.; Saliu, F.; Lasagni, M.; Ghezzi, L.L.M.; Balestrieri, A. Microplastic contamination in terrestrial ecosystems: A study using barn owl (*Tyto alba*) pellets. *Chemosphere* **2022**, *308 Pt 1*, 136281. [CrossRef]
38. Milchev, B.; Chobanov, D.; Simov, N. Diet and foraging habitats of non-breeding white storks (*Ciconia ciconia*) in Bulgaria. *Arch. Biol. Sci.* **2013**, *65*, 1007–1013. [CrossRef]
39. Reynolds, C.; Ryan, P.G. Micro-plastic ingestion by waterbirds from contaminated wetlands in South Africa. *Mar. Pollut. Bull.* **2018**, *126*, 330–333. [CrossRef]
40. Carlin, J.; Craig, C.; Little, S.; Donnelly, M.; Fox, D.; Zhai, L.; Walters, L. Microplastic accumulation in the gastrointestinal tracts in birds of prey in central Florida, USA. *Environ. Pollut.* **2020**, *264*, 114633. [CrossRef] [PubMed]
41. Andrady, A.; Barnes, P.; Bornman, J.; Gouin, T.; Madronich, S.; White, C.; Zepp, R.; Jansen, M. Oxidation and fragmentation of plastics in a changing environment; from UV-radiation to biological degradation. *Sci. Total Environ.* **2022**, *851*, 158022. [CrossRef] [PubMed]

42. Welsh, B.; Aherne, J.; Paterson, A.M.; Yao, H.; McConnell, C. Atmospheric deposition of anthropogenic particles and microplastics in south-central Ontario, Canada. *Sci. Total Environ.* **2022**, *835*, 155426. [CrossRef]
43. Moore, C.J. Synthetic polymers in the marine environment: A rapidly increasing, long-term threat. *Environ. Res.* **2008**, *108*, 131–139. [CrossRef] [PubMed]
44. Lundquist, L.; Leterrier, Y.; Sunderland, P.; Månson, J. *Life Cycle Engineering of Plastics: Technology, Economy and Environment*; Elsevier: Amsterdam, The Netherlands, 2001.
45. Abraham, J.; Ghosh, E.; Mukherjee, P.; Gajendiran, A. Microbial degradation of low density polyethylene. *Environ. Prog. Sustain. Energy* **2017**, *36*, 147–154. [CrossRef]
46. Sarker, M.; Mamunor Rashid, M.; Sadikur Rahman, M. Low Density Polyethylene (LDPE) Waste Plastic Transformation into Renewable Heavy Fuel Using Thermal Cracking. *World Environ.* **2012**, *2*, 140–147. [CrossRef]
47. Sudirman; Dharmayani, N.K.T.; Yuanita, E.; Ulfa, M.; Sudarma, I.M.; Ikhsan, A. Reprocessing plastic waste into petroleum fraction based on zero waste principle. *AIP Conf. Proc.* **2020**, *2243*, 020026.
48. Habschied, K.; Mastanjević, K.; Šibalić, M.; Krstanović, V.; Galić, V. A Survey on Detection of Plastic-Related Chemicals in Beer Packaged in PET Using FT-IR Technology. *Beverages* **2022**, *8*, 73. [CrossRef]
49. Bocqué, M.; Lapinte, V.; Courault, V.; Couve, J.; Cassagnau, P.; Robin, J. Phosphonated Lipids as Primary Plasticizers for PVC with Improved Flame Retardancy. *Eur. J. Lipid Sci. Technol.* **2018**, *120*, 1800062. [CrossRef]
50. Barlow, C.Y.; Morgan, D.C. Polymer film packaging for food: An environmental assessment. *Resour. Conserv. Recycl.* **2013**, *78*, 74–80. [CrossRef]
51. Katsura, T.; Sasaki, H. On-going solutions to environmental issues in plastic packaging. *Packag. Technol. Sci.* **2001**, *14*, 87–95. [CrossRef]
52. Mattetti, M.; Beltramin, A.; Perez Estevez, M.A.; Varani, M.; Renzi, M.; Alberti, L. Start and stop systems on agricultural tractors as solution for saving fuel and emissions. *Biosyst. Eng.* **2022**, *216*, 108–120. [CrossRef]
53. de Boer, C.; Paraffins, C.; Boer, J. (Eds.) *The Handbook of Environmental Chemistry*; Springer: Berlin/Heidelberg, Germany, 2010; Volume 10.
54. Espinosa, C.; Esteban, M.Á.; Cuesta, A. Microplastics in Aquatic Environments and Their Toxicological Implications for Fish. In *Toxicology—New Aspects to This Scientific Conundrum*; Intechopen: London, UK, 2016.
55. Wu, X.; Chen, X.; Jiang, R.; You, J.; Ouyang, G. New insights into the photo-degraded polystyrene microplastic: Effect on the release of volatile organic compounds. *J. Hazard. Mater.* **2022**, *431*, 128523. [CrossRef]
56. Liu, H.-D.; Gu, B.; Jia, F.; Li, Y.; Ying, Q.; Alamus; Yuan, W.F.; Chen, B.; He, Q. Facile fabrication of hydrophobic octadecylamine-functionalized polyurethane foam for oil spill cleanup. *J. Macromol. Sci. Part A* **2016**, *53*, 196–200. [CrossRef]
57. Bohrer, D. *Sources of Contamination in Medicinal Products and Medical Devices*; John Wiley & Sons: Hoboken, NJ, USA, 2012.
58. Narciso-Ortiz, L.; Coreño-Alonso, A.; Mendoza-Olivares, D.; Lucho-Constantino, C.A.; Lizardi-Jiménez, M.A. Baseline for plastic and hydrocarbon pollution of rivers, reefs, and sediment on beaches in Veracruz State, México, and a proposal for bioremediation. *Environ. Sci. Pollut. Res. Int.* **2020**, *27*, 23035–23047. [CrossRef]
59. Varjani, S.J. Microbial degradation of petroleum hydrocarbons. *Bioresour. Technol.* **2017**, *223*, 277–286. [CrossRef]
60. García-Cruz, N.; Valdivia-Rivera, S.; Narciso-Ortiz, L.; García-Maldonado, J.; Uribe-Flores, M.; Aguirre-Macedo, M.; Lizardi-Jiménez, M. Diesel uptake by an indigenous microbial consortium isolated from sediments of the Southern Gulf of Mexico: Emulsion characterisation. *Environ. Pollut.* **2019**, *250*, 849–855. [CrossRef] [PubMed]
61. Nizzetto, L.; Futter, M.; Langaas, S. Are Agricultural Soils Dumps for Microplastics of Urban Origin? *Environ. Sci. Technol.* **2016**, *50*, 10777–10779. [CrossRef] [PubMed]
62. Souza Machado, A.A.; Kloas, W.; Zarfl, C.; Hempel, S.; Rillig, M.C. Microplastics as an emerging threat to terrestrial ecosystems. *Glob. Chang. Biol.* **2018**, *24*, 1405–1416. [CrossRef] [PubMed]
63. Kumar, M.; Xiong, X.; He, M.; Tsang, D.C.; Gupta, J.; Khan, E.; Harrad, S.; Hou, D.; Ok, Y.S.; Bolan, N.S. Microplastics as pollutants in agricultural soils. *Environ. Pollut.* **2020**, *265*, 114980. [CrossRef] [PubMed]
64. Yukioka, S.; Tanaka, S.; Nabetani, Y.; Suzuki, Y.; Ushijima, T.; Fujii, S.; Takada, H.; Van Tran, Q.; Singh, S. Occurrence and characteristics of microplastics in surface road dust in Kusatsu (Japan), Da Nang (Vietnam), and Kathmandu (Nepal). *Environ. Pollut.* **2020**, *256*, 113447. [CrossRef] [PubMed]
65. Edo, C.; González-Pleiter, M.; Leganés, F.; Fernández-Piñas, F.; Rosal, R. Fate of microplastics in wastewater treatment plants and their environmental dispersion with effluent and sludge. *Environ. Pollut.* **2020**, *259*, 113837. [CrossRef] [PubMed]
66. Hancock, J.; Kushlan, J.; Kahl, M. *Storks, Ibises and Spoonbills of the World*; Princeton University Press: Princeton, NJ, USA, 2010.
67. Alonso, J.C.; Alonso, J.A.; Carrascal, L.M. Habitat selection by foraging White Storks, *Ciconia ciconia*, during the breeding season. *Can. J. Zool.* **1991**, *69*, 1957–1962. [CrossRef]
68. Vrezec, A. Insects in the White Stork *Ciconia ciconia* diet as indicators of its feeding conditions: The first diet study in Slovenia. *Acrocephalus* **2009**, *30*, 25–29. [CrossRef]
69. Johst, K.; Brandl, R.; Pfeifer, R. Foraging in a Patchy and Dynamic Landscape: Human Land Use and the White Stork. *Ecol. Appl.* **2001**, *11*, 60. [CrossRef]
70. Kwieciński, Z.; Kwiecińska, H.; Ratajszczak, R.; Ćwiertnia, P.; Tryjanowski, P. Food selection of the white storks *Ciconia ciconia* under captive conditions. In *White Stork Study in Poland: Biology, Ecology and Conservation*; Bogucki Wydawnictwo Naukowe: Poznań, Poland, 2015.

71. Mužinić, J.; Rašajski, J. On food and feeding habits of the White Stork, *Ciconia cciconia*, in Central Balkans. *Okol. Der Vögel* **1992**, *14*, 211–223.
72. Rékási, J. A study of the White Stork population of North Bácska in 1999. *Ornis Hungarica* **2000**, *10*, 225–229.
73. Surdo, S.; Zapparrata, C.; Ientile, R.; Massa, B. Evidence suggests an opportunistic entomophagous diet of the White Stork *Ciconia ciconia* in Sicily during breeding and post-breeding periods. *Avocetta* **2022**, *46*, 49–56.

toxics

MDPI

Article

The Exploration of Joint Toxicity and Associated Mechanisms of Primary Microplastics and Methamphetamine in Zebrafish Larvae

Hao Wang [1], Jindong Xu [1], Yang Yuan [1], Zhenglu Wang [2], Wenjing Zhang [1] and Jiana Li [3,*]

[1] College of Oceanography, Hohai University, Nanjing 210098, China; 15725388264@163.com (H.W.); aukuma@gmail.com (J.X.); yy15252540515@126.com (Y.Y.); 15854287023@163.com (W.Z.)

[2] West China School of Public Health, West China Fourth Hospital Sichuan University, Chengdu 610041, China; m18018567955@163.com

[3] Ningbo Academy of Ecological, Environmental Sciences, Ningbo 315000, China

* Correspondence: stlijiana@163.com; Tel.: +86-185-1258-1998

Abstract: The co-existence of microplastics (MPs) and methamphetamine (METH) in aquatic ecosystems has been widely reported; however, the joint toxicity and associated mechanisms remain unclear. Here, zebrafish larvae were exposed individually or jointly to polystyrene (PS) and polyvinyl chloride (PVC) MPs (20 mg/L) and METH (1 and 5 mg/L) for 10 days. The mortality, behavioral functions, and histopathology of fish from different groups were determined. PS MPs posed a stronger lethal risk to fish than PVC MPs, while the addition of METH at 5 mg/L significantly increased mortality. Obvious deposition of MPs was observed in the larvae's intestinal tract in the exposure groups. Meanwhile, treatment with MPs induced intestinal deposits and intestinal hydrops in the fish, and this effect was enhanced with the addition of METH. Furthermore, MPs significantly suppressed the locomotor activation of zebrafish larvae, showing extended immobility duration and lower velocity. METH stimulated the outcome of PS but had no effect on the fish exposed to PVC. However, combined exposure to MPs and METH significantly increased the turn angle, which declined in individual MP exposure groups. RNA sequencing and gene quantitative analysis demonstrated that exposure to PS MPs and METH activated the MAPK signaling pathway and the C-type lectin signaling pathway of fish, while joint exposure to PVC MPs and METH stimulated steroid hormone synthesis pathways and the C-type lectin signaling pathway in zebrafish, contributing to cellular apoptosis and immune responses. This study contributes to the understanding of the joint toxicity of microplastics and pharmaceuticals to zebrafish, highlighting the significance of mitigating microplastic pollution to preserve the health of aquatic organisms and human beings.

Keywords: microplastics; methamphetamine; joint toxicity; zebrafish; health

1. Introduction

Microplastics (MPs) typically refer to plastic particles with a size (longest dimension of the particle) of less than 5 mm, which take the form of microspheres, fibers, and fragments [1]. Considered an emerging pollutant, microplastics undergo a succession of physical, chemical, and biological processes, ultimately permeating the environment, including rivers, oceans, lakes, and soil [2,3]. Due to their wide distribution, resistance to degradation, affinity for adsorbing environmental pollutants, and ease of entering organisms, the ubiquity of microplastics in aquatic ecosystems might pose a threat to both ecological and human health [4,5]. Some scholars examined the microplastic pollution in Turkey's water environments and found that there was microplastic pollution in all kinds of these environments in the country [6]. Given their small size, MPs can be mistaken for food and ingested by zooplankton, fish, and crustaceans [7]. Hence, microplastics have been detected within various organisms, including mammals, birds, mollusks, and fish [8,9], further indicating the associated environmental risks. Previous studies have shown that exposure to microplastics could induce physical damage, inhibit physiological and biochemical functions, alter behavioral patterns, and reduce reproductive capacity and survival rates of organisms [10,11]. Studies have shown that rainbow trout exposed to polyethylene can exhibit growth inhibition, tissue damage, and cell apoptosis [12]. In addition, ingestion of microplastics by fish can lead to tissue oxidative stress and inhibit enzyme activity [13]. At present, controlling microplastic pollution is very urgent. Research has found that commercial feed provided during fish growth can also lead to a certain degree of microplastic pollution in various fish products [14]. Among the different types of MPs, the predominant ones in the environment were polypropylene (PP), polystyrene (PS), and polyvinyl chloride (PVC) [15]. PVC is one of the most produced commodity plastics globally, being extensively used in various human activities [16]. Some studies have also shown that PVC exhibits certain acute toxicity, as large copepods exposed to PVC leachates displayed signs of toxicity [17]. African catfish (*Clarias galliepinus*) exposed to PVC particles can exhibit adverse reactions such as neurotoxicity and oxidative stress [18]. Similarly, PS is one of the most commonly produced plastics globally [19], with research revealing that PS microbeads can accumulate in sea urchin intestines and negatively affect their embryonic development [20]. Research has also found that ingestion of PS particles can cause damage to the liver and gills of rainbow trout, resulting in negative impacts on their health [21]. Scholars have also found that polystyrene microplastics can weaken the predator-induced defenses of *Daphnia*, which may have a certain impact on aquatic ecology [22]. Although many studies have investigated the toxicity of PS or PVC to aquatic animals, the differences in toxicity induced in aquatic species by PS and PVC are so far unknown.

Besides the effect of single MPs, it is well known that MPs have a strong absorption capacity for organic chemicals [23]. Therefore, the co-existence of MPs and pollutants creates combined toxicity to organisms [24]. For example, MPs antagonized the toxicity of microalgae induced by triclosan [25] while enhancing the adverse effects on algae caused by procainamide [26]. A study has found that when polyethylene microplastics are combined with acetochlor, PE MPs significantly enhance the acute toxicity of ACT to zebrafish. In addition, it also increases the accumulation of ACT in zebrafish and exacerbates the oxidative stress damage of ACT in the intestine [27]. Scholars have conducted joint exposure of microplastics and phenanthrene to marine medaka (*Oryzias melastigma*), and the results indicate that the interaction between MPs and Phe may have a significant impact on the gut microbiota and metabolism of aquatic organisms [28]. Hence, the combined toxicity of MPs and emerging organic pollutants on aquatic organisms should be a concern.

Methamphetamine (METH) is a synthetic drug commonly used as a stimulant and hallucinogen [29]. Recently, it has become a prevalent illicit drug due to its highly addictive nature [30]. Similar to traditional pharmaceuticals, METH cannot be metabolized completely and thus enters the environment in the form of parent and metabolites [31]. The toxic effects and dependency mechanisms of METH are associated with central nervous system signaling pathways, the dopaminergic system, neurotransmitters, and molecular genetics [32]. Considering the conservation of the central nervous signaling pathways between vertebrates, exposure to METH might interact with dopamine, norepinephrine, and serotonin receptors in fish [32], showing a stimulating response [33]. Additionally, it has been revealed that METH can induce neuronal apoptosis, autophagy, and behavioral sensitization [34–36], causing oxidative stress and neuroinflammation [37,38]. The presence of METH in water environments might pose hazards to aquatic organisms and ecosystems. For example, research has indicated that the presence of METH in water interferes with the respiration, metabolism, and reproductive functions of aquatic organisms, leading to death and growth retardation [39]. The ecology-associated behaviors and processes of aquatic organisms (i.e., elegans) were disrupted by METH at low levels (50 ng/L), showing ecological system outcomes [40]. A previous study has found that the adsorption of METH by microplastics could lead to increased METH ingestion by aquatic organisms [41]. Therefore, the combined effects of METH with MPs in aquatic environments are highly likely to pose an increased threat to aquatic organisms.

This study aims to elucidate the combined toxicity of PVC and PS with METH by using zebrafish as an animal model. Zebrafish was selected as the model organism because it is a well-established model for assessing the toxicity of various substances [42]. The indicators, including mortality, behavioral functions, and histopathology of zebrafish in different groups, were determined. Furthermore, transcriptomic profiles of fish exposed to MPs and MPs combined with METH were established, and the enriched genes were quantitatively analyzed. Accordingly, the underlying toxicity mechanisms were elucidated.

2. Materials and Methods

2.1. Experimental Animals and Materials

Zebrafish embryos were obtained from Shanghai FeiXi Biotechnology Co., Ltd. (Shanghai, China) and incubated at the Marine Science Experimental Center, Hohai University. The animal research protocol was approved by the Institutional Animal Care and Use Committee. PVC and PS MPs powders (average diameter 10 μm) were purchased from Dongguan Junxin Plastics Co., Ltd. (Dongguan, China). The purity and characteristics of microplastics powders were identified by using a Thermo Scientific Nicolet iN10 infrared spectrometer (Waltham, MA, USA) (Figure S1), and particle size distribution was measured by using a Malvern Mastersizer 2000 laser particle size analyzer (Akribis Scientific Limited, Cheshire, UK) (Figure S2).

Zebrafish embryos (24 h post fertilization) were randomly placed in a fresh 60 mm Petri dish (20 each) spiked with embryo-rearing media (ERM; containing H_2O, NaCl, $CaCl_2$, KCl, $MgSO_4$, pH 7.2). After hatching, zebrafish larvae were transferred to new Petri dishes for further cultivation, maintaining strict water quality and environmental conditions with dissolved oxygen > 7.0 mg/L, pH 7.0~7.4, no residual chlorine, ammonia, or nitrite. The light cycle was set at 14:10, the cultivation temperature was maintained at 26 ± 1 °C, and the larvae were fed twice daily with fresh brine shrimp nauplii. METH hydrochloride was purchased from Cerilliant Corporation (Round Rock, TX, USA), and a stock solution (1 tmg/mL) was prepared by using deionized water and diluted with ERM as needed. Three replicates for one exposure concentration were set: (1) control; (2) 20 mg/L PS MPs; (3) 20 mg/L PS MPs + 1 mg/L METH (PS + 1); (4) 20 mg/L PS MPs + 5 mg/L METH (PS + 5); (5) 20 mg/L PVC MPs; (6) 20 mg/L PVC MPs + 1 mg/L METH (PVC + 1); (7) 20 mg/L PVC MPs + 5 mg/L METH (PVC + 5). The concentrations we used were obtained from other acute exposure experiments in the literature [41,43–47]. The exposure experiment lasted for 10 days with daily renewal of the exposure solutions in each Petri dish.

2.2. Survival Rate Determination

During the exposure period, the morphology of each zebrafish larvae was daily observed by using a stereo microscope (Upshine Instrument Co., Ltd., Suzhou, China). The basis for determining the death of zebrafish juveniles was the observation of cardiac arrest under stereo microscope [48]. Survival and mortality rates of zebrafish larvae in each experimental group were recorded daily.

2.3. Behavioral Assessment

At the end of exposure, behavioral functions of zebrafish larvae were assayed (eleven replicates for each group, n = 11) according to a previous publication [49]. Briefly, the locomotor trajectory of zebrafish was recorded using a CCD camera installed at the top of the tank for a duration of 10 min after 30 min acclimation, then analyzed by using XT7 software (Noldus IT, Wageningen, The Netherlands) to obtain the total swimming data of fish [50]. When observing the movement trajectories of zebrafish larvae, each extracted larva was individually observed. The behavioral functional indicators analyzed were mean velocity (cm/s), total distance (cm), turn angle, and immobility duration (%).

2.4. Histopathological Examination

The zebrafish were sacrificed by being rapidly cooled down after the 10-day exposure. Considering the small size of the animal, the whole of each larva was fixed using 4% paraformaldehyde (PFA) solution for 24 h, followed by dehydration using a gradation of ethanol series (increasing concentrations from high to low), and then embedded in paraffin (five replicates for each group, n = 5). Cross sections of approximately 8 μm thickness were obtained from the paraffin blocks and stained with hematoxylin and eosin (H&E). The obtained tissue sections were analyzed for histopathological changes by using an optical microscope.

2.5. RNA Extraction, Sequencing, and Analysis

In order to identify pathways and genes affected by combined toxicity and better comply with the 3R principle, the total RNA of the zebrafish larvae in the control, PS, PS + 5, PVC, and PVC + 5 groups was extracted by Trizol (Invitrogen, Carlsbad, CA, USA) according to the manufacturer's protocol (three replicates for each group). RNA purity was checked by using the NanoPhotometer® spectrophotometer (IMPLEN, Calabasas, CA, USA), integrity was assessed by using the RNA Nano 6000 Assay Kit of the Agilent Bioanalyzer 2100 system (Agilent Technologies, Santa Clara, CA, USA), and concentration was measured by using the Qubit® RNA Assay Kit in the Qubit®2.0 Fluorometer (Life Technologies, Carlsbad, CA, USA). Library construction and RNA sequencing were performed on an Illumina Novaseq platform. The detailed protocols are shown in the Supplementary Materials.

2.6. Quantitative Real-Time PCR Analysis

According to the results of transcriptomics analysis, seven genes (i.e., *srfa*, *elk1*, *elk2*, *jun*, *mapk9*, *nfatc2a*, and *nfatc4*) of the fish from the control, PS, and PS + 5 groups were relatively quantitatively analyzed, while six genes (i.e., *cyp3c1*, *cyp3c4*, *cyp19a1a*, *pla2g4ab*, *nfatc2a*, and *nfatc4*) of the fish from the control, PVC, and PVC + 5 groups were chosen as candidates for Q-RT-PCR analysis. The detailed protocols are described in the Supplementary Materials, and all primer sequences are shown in Table S1.

2.7. Statistical Analysis

The statistical program SPSS 18.0 (Chicago, IL, USA) was used to analyze all the collected data. Five or more replicates of each parameter were determined to eliminate the variability of the results. All the data are expressed as mean $\pm$ standard deviations (S.D.). One-way analysis of variance (ANOVA) was used to compare and analyze the differences in mortality rate and behavioral parameters. One-way ANOVA followed by Turkey's test was used to determine the differences in gene transcription levels of zebrafish larvae between the exposure and control groups. Two-way ANOVA analysis was used to determine the differences in the temporal changes of mortality of fish in the different groups. The differences in the mortality rates of fish in different groups were calculated by using the χ^2 test. A *p*-value of less than 0.05 (95% confidence interval) was considered statistically significant.

3. Results

3.1. The Mortality of Zebrafish Larvae Was Influenced by MPs or MPs and METH, and Histopathological Changes Occured in the Tissues

There was no mortality observed among zebrafish larvae in both the control and exposure groups during the acclimation stage. As shown in Figure 1a, the mortality of fish in the PS exposure groups was significantly greater than in the control from the sixth day onward ($p < 0.01$), and the climbing speed in the PS + 5 group was much higher than in the other exposure groups. Similarly, the mortality in the PS groups was significantly greater than in the control from the fifth day, and the patterns of change in the PVC and PVC + 5 groups were parallel, showing steeper increases than in the PVC + 1 group (see Figure 1b).

For the final mortality rates of different groups (Figure 1c,d), the values in the PS, PS + 1, and PS + 5 groups (33.3%, 30%, and 43.3%, respectively) were much higher than in the control (10%, $p < 0.05$), with the peak observed in the PS + 5 group (significantly greater than in the PS and PS + 1 groups). The absolute levels in the PVC (30%), PVC + 1 (26.7%), and PVC + 5 (30%) groups were markedly greater than in the control (10%, $p < 0.05$), with the peak in the PVC group (insignificant differences among the exposure groups).

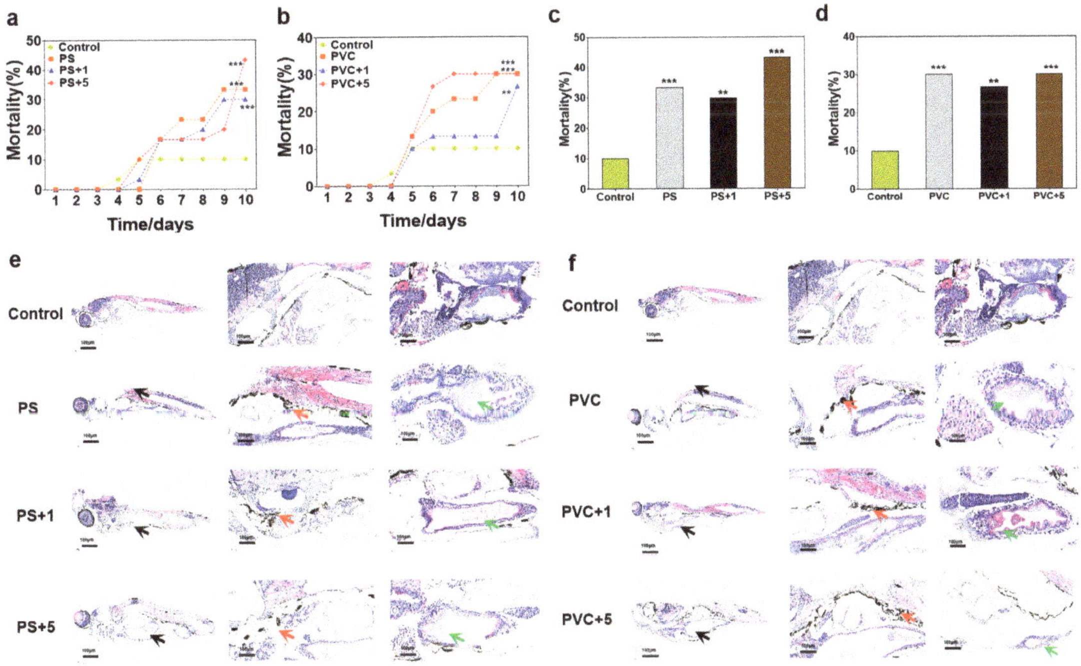

Figure 1. The mortality of zebrafish larvae was influenced by MPs or MPs and METH, and histopathological changes occurred in the tissues. The trajectory curves of mortality at different time points of PS, PS + 1, and PS + 5 (**a**). The trajectory curves of mortality at different time points of PVC, PVC + 1, and PVC + 5 (**b**). The terminal mortality of control, PS, PS + 1, and PS + 5 (**c**). The terminal mortality of control, PVC, PVC + 1, and PVC + 5 (**d**). ** $p < 0.01$, *** $p < 0.001$ vs. the control. Histopathological effects of zebrafish larvae after 10 days of exposure to MPs and METH (**e**,**f**). Abnormal morphology of enlarged trunk (black arrow), intestinal deposits (red arrows), and intestinal hydrops (green arrow).

According to the results of H&E staining of the whole body, obvious MP deposition and intestinal hydronephrosis were found in the fish in the exposure groups compared to the control (where the larvae displayed normal morphology, and no significant abnormalities were observed in the internal tissues). As shown in Figure 1e, the larvae in the PS exposure groups showed abnormal morphology compared to the control (black arrow), the front halves of juvenile fish trunks were found to become enlarged in the combined exposure groups. There were obvious particle depositions in the gastrointestinal tracts of fish (red arrow) in the exposure groups, with insignificant differences between the different PS groups. Moreover, an inflated enteric cavity was observed in the larvae from the exposure groups compared to the control (green arrow), indicating the occurrence of intestinal hydrops, showing a dose-dependent response. The changes in the PVC exposure groups showed similar patterns to the PS groups (Figure 1f).

3.2. The Behavioral Functions Were Affected by MPs or MPs and METH

Through analysis of recorded movement trajectories of zebrafish larvae, it was observed that larvae in the control group exhibited exploratory swimming behavior characterized by long distances swam and strong randomness in swimming direction (Figure 2a). However, individual or combined exposure of MPs and METH significantly limited the locomotion of the larvae, which showed relatively simple swimming patterns in comparison with the control (Figure 2a). Statistically (Figure 2b,c), the immobility (%) of fish in the PS groups (51%, 39.5%, and 22.8%) was significantly higher than in the control group (3.2%). The combined exposure of PS and METH partially neutralized this effect, and the value of the PS + 5 group was significantly lower than that of the PS exposure group. Moreover, the mean velocity (cm/s) of the fish in the PS groups (55.3 cm/s, 86.7 cm/s, and 36.3 cm/s) declined markedly compared to the control group (106.9 cm/s), and there were no differences among the exposure groups. The individual exposure to PS decreased the turn angle of the larvae ($p < 0.05$), and the addition of METH ameliorated this change. For PVC groups, the immobility (%) of fish in the exposure groups (74.1%, 67.8%, and 77.1%) were markedly greater than the control group (3.2%), while the mean velocity (cm/sec) values were much lower (9.8 cm/s, 11.1 cm/s, and 14.3 cm/s), and no differences were found between different exposure groups. Notably, PVC individual treatment decreased the turn angle (118.9), but the addition of METH increased the turn angle of the fish, for which the value in the PVC + 5 (255.4) group was markedly higher than that in the control group (185.5).

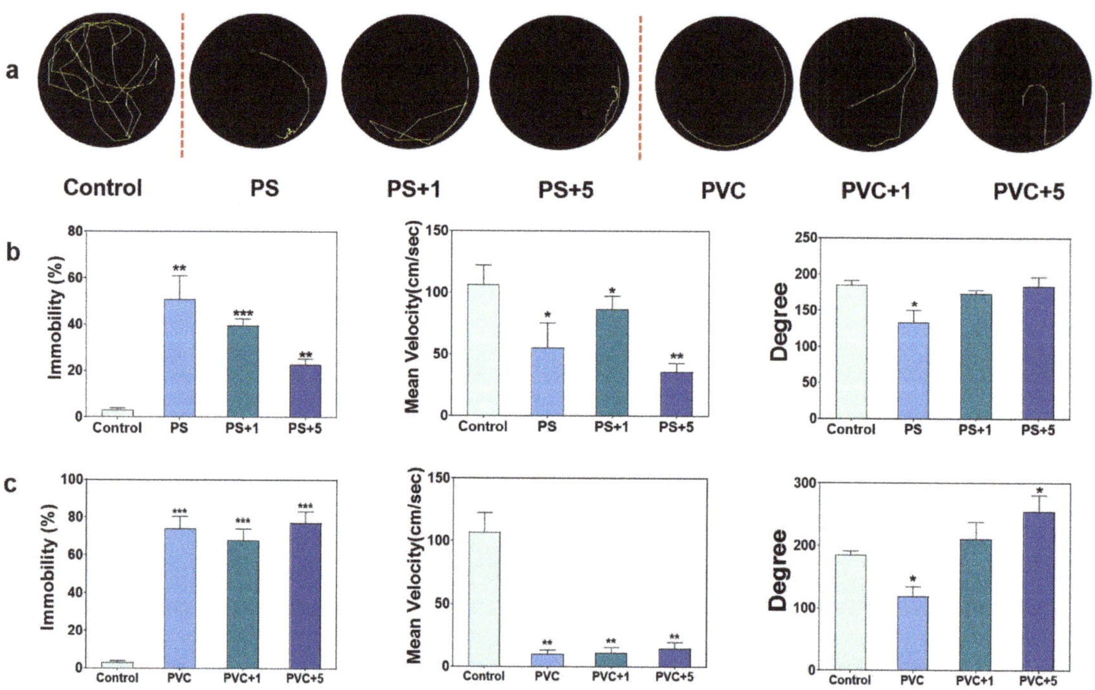

Figure 2. Behavioral changes in zebrafish larvae after 10 days of exposure to MPs and METH. Behavioral trajectories of zebrafish larvae after exposure, photographed from above (**a**). The mobility rate, mean velocity, and degree of zebrafish larvae of the group exposed to PS and METH (**b**). The mobility rate, mean velocity, and degree of zebrafish larvae of the group exposed to PVC and METH (**c**). * $p < 0.05$, ** $p < 0.01$, *** $p < 0.001$ vs. the control.

3.3. The Gene Profiles of Zebrafish Larvae Were Affected by MPs or MPs and METH

To elucidate the underlying molecular mechanisms of the adverse outcomes of zebrafish larvae induced by MPs or MPs + METH, RNA-seq was performed to establish the whole-genome expression profiling changes of fish in the control, PS, PS + METH, PVC, and PVC + METH groups. For the control and PS exposure group (Figure 3a), 551 differentially expressed genes (DEGs) were identified, comprising 226 upregulated genes and 325 downregulated genes, while there were 597 DEGs identified between the control and PS + 5 groups (175 upregulation and 422 downregulation, Figure 3b). Between the PS experimental group and the PS + 5 experimental group (Figure 3c), 235 DEGs were enriched, with 110 genes upregulated and 125 genes downregulated.

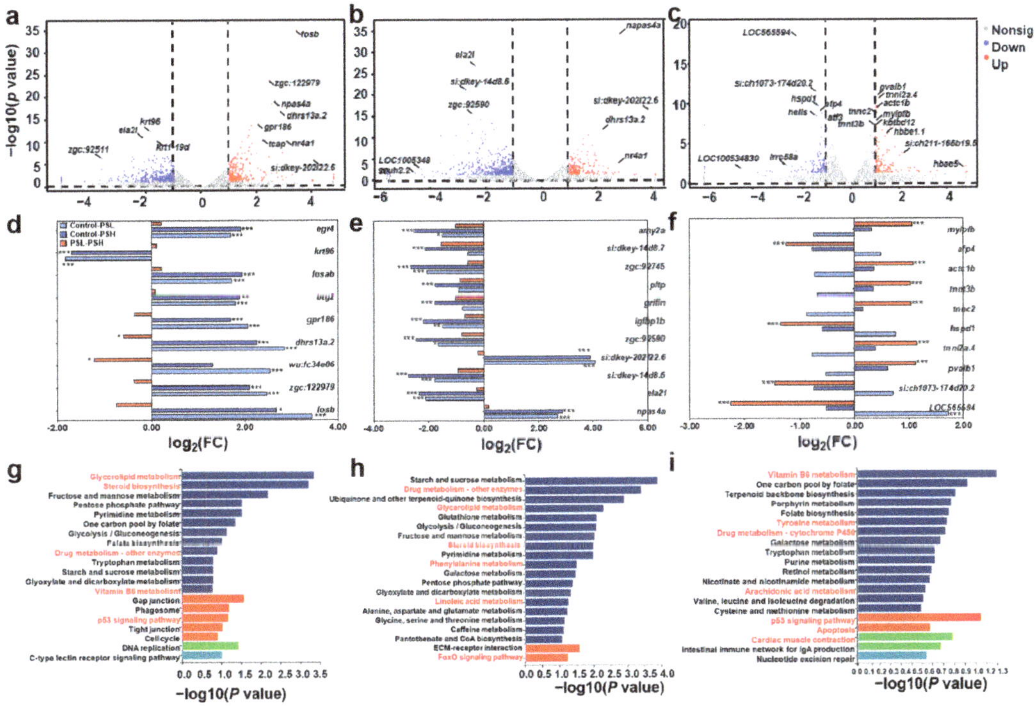

Figure 3. The gene profiles of zebrafish larvae were affected by PS or PS and METH. The differentially expressed genes (**a**–**c**). The top 30 DEGs (**d**–**f**). * $p < 0.05$, ** $p < 0.01$, *** $p < 0.001$. The enrichment pathways (**g**–**i**).

Among the DEGs, there were some overlap genes, like the downregulated gene ela2l and the upregulated gene napas4a. For the top 30 DEGs (Figure 3d–f), most genes from the control–PS and control–PS + 5 groups changed with consistent patterns, but they were different from the PS–PS + 5 groups, like *egr4*, *fosab*, *fosb*, *ela2l*, *npas4a*, and *btg2*. As a result, different pathways were enriched (Figure 3g–i). Between the control and exposure groups, drug metabolism and the P53 signaling pathway were enriched. For control–PS, glycerolipid metabolism, steroid biosynthesis, and vitamin B6 metabolism were significantly identified, while phenylalanine metabolism, linoleic acid metabolism, and the foxO signaling pathway were identified from the control–PS + 5 groups. Between the PS and PS + 5 groups, tyrosine metabolism, arachidonic acid metabolism, and cardiac muscle contraction were predominant. Based on the DEGs and the enrichment pathways, the correlation networks were established (Figure S3). The downregulated genes *cel.1* and *cel.2* were identified both in glycerolipid metabolism and steroid biosynthesis from the control–PS and control–PS + 5 groups, while the upregulated genes *rrm2* and *LOC100330664* involved in the P53 signaling pathway and drug metabolism were annotated. Meanwhile, the upregulated gene *plk2b* was linked to the FoxO signaling pathway. However, the genes involved in the P53 signaling pathway and drug metabolism were different from the PS and PS + 5 groups, *cycsb* and *aox5*, respectively.

For the control and PVC exposure groups (Figure 4a), 427 differentially expressed genes (DEGs) were identified, comprising 285 upregulated genes and 142 downregulated genes, while there were 322 DEGs identified between the control and PVC + 5 groups (33 upregulation and 289 downregulation, Figure 4b). Between the PVC experimental group and the PVC + 5 experimental group (Figure 4c), 661 DEGs were enriched, with 51 genes upregulated and 610 genes downregulated.

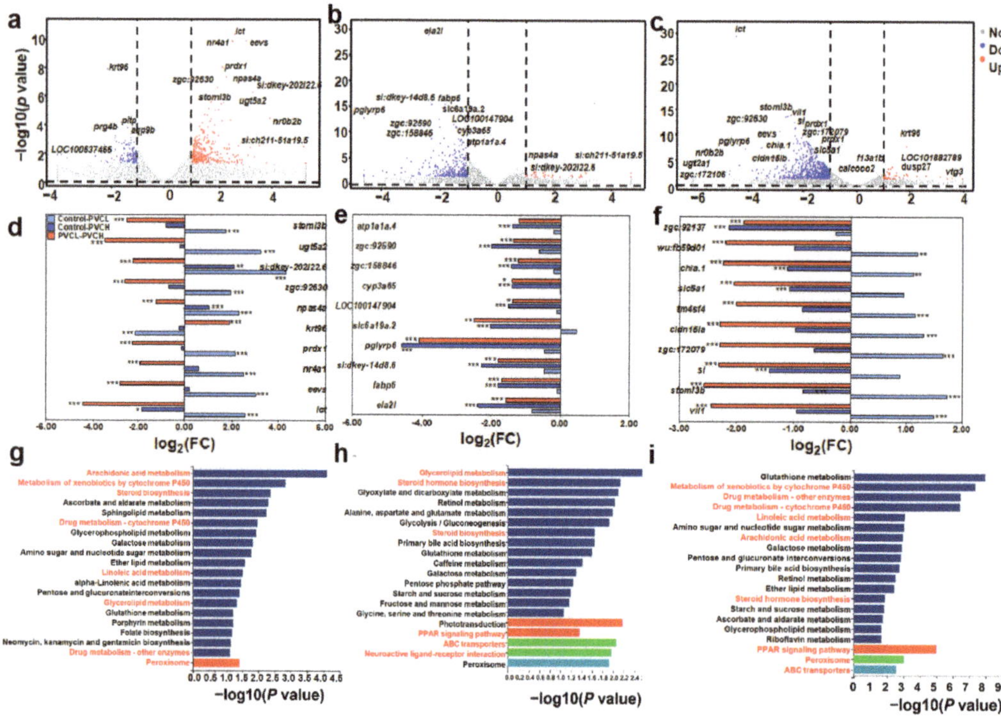

Figure 4. The gene profiles of zebrafish larvae were affected by PVC or PVC and METH. The differentially expressed genes (**a**–**c**).The top 30 DEGs (**d**–**f**). * $p < 0.05$, ** $p < 0.01$, *** $p < 0.001$. The enrichment pathways (**g**–**i**).

Among the DEGs, there were some overlap genes, like the downregulated gene fabp6 and the upregulated gene napas4a. For the top 30 DEGs (Figure 4d–f), most genes from the control–PVC + 5 and PVC–PVC + 5 groups changed with consistent patterns, but they were different from the control–PVC groups, like *lct*, *cyp3a65*, *plyrp6*, *fabp6*, *npas4a*, and *si*. As a result, different pathways were enriched (Figure 4g–i). Between the control and exposure groups, glycerolipid metabolism and steroid biosynthesis were enriched. For the control–PVC groups, arachidonic acid metabolism, metabolism of xenobiotics by cytochrome P450, and peroxisome were significant pathways, while steroid hormone biosynthesis, the PPAR signaling pathway, and ABC transporter were identified from the control–PVC-5 group. Between the PVC and PVC-5 groups, metabolism of xenobiotics by cytochrome P450, linoleic acid metabolism, and peroxisome were predominant. Based on the DEGs and the enrichment pathways, the correlation networks were established

(Figure S4). The downregulation gene *acsl5* was identified both in the PPAR signaling pathway and peroxisome from the control–PVC + 5 and PVC–PVC + 5 groups. Meanwhile, the upregulated gene *abcg8* was linked to ABC transporter, and the downregulated gene *agxta* was linked to peroxisome. However, the genes involved in the peroxisome were different from those in the control–PVC group, like *ech1* and *pradx1*.

3.4. The Gene Expression Levels of Zebrafish Larvae Affected by MPs or MPs and METH

Based on the RNA sequencing results of zebrafish larvae, the genes *srfa*, *elk1*, and *elk2* involved in the MAPK signaling pathway, *jun* and *mapk9* involved in apoptosis pathways, and *nfatc2a* and *nfatc4* involved in the C-type lectin signaling pathway in the PS groups were quantitatively analyzed (Figure 5a). Genes *srfa*, *elk1*, and *elk2* were significantly upregulated by PS + METH at 5 mg/L ($p < 0.05$) with 6.1-fold, 13.8-fold, and 55.8-fold, respectively, while individual exposure to PS showed no effects on the expression levels ($p > 0.05$ vs. the control). However, the significant upregulation of gene *jun* was found both in the PS and PS + 5 groups ($p < 0.05$, 15.8-fold and 15.4-fold). The increased transcriptional levels of *mapk9* were shown in the PS and PS + 5 groups with dose–response change ($p < 0.05$, 3.8-fold and 7.9-fold). Meanwhile, the expression levels of the genes *nfatc2a* and *nfatc4* were stimulated by combined exposure of PS and METH at 5 mg/L (4.6-fold and 26.7-fold, respectively).

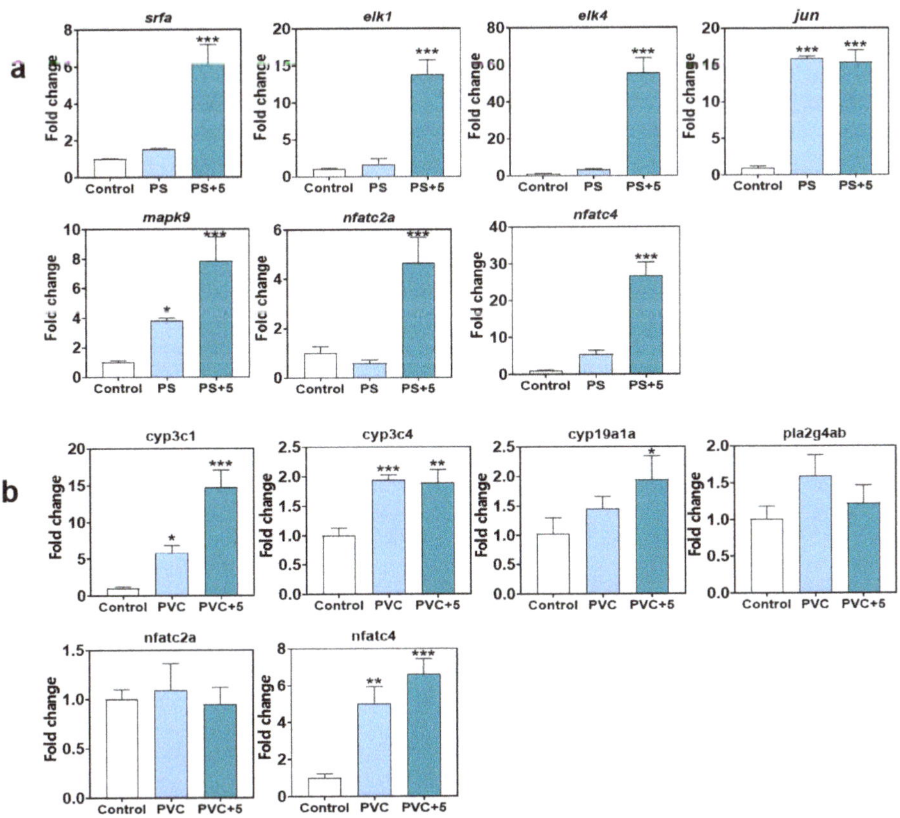

Figure 5. The gene expression levels of zebrafish larvae affected by MPs or MPs and METH. Gene expression levels of zebrafish juveniles in the PS groups (**a**). Gene expression levels of zebrafish juveniles in the PVC groups (**b**). All the data are expressed as mean ± S.D. * $p < 0.05$, ** $p < 0.01$, *** $p < 0.001$ vs. controls; post hoc Tukey's test for significant ANOVA data.

Meanwhile, the genes *cyp3c1*, *cyp3c4*, and *cyp19a1a* (involved in steroid hormone biosynthesis pathways), *pla2g4ab* (involved in the glycerophospholipid metabolism pathway), and *nfatc2a* and *nfatc4* (involved in the C-type lectin signaling pathway) of zebrafish larvae in PVC groups were quantitatively analyzed (Figure 5a). Genes, *cyp3c1*, *cyp3c4*, and *cyp19a1a* were all upregulated in the PVC and PVC + 5 groups, and were significantly upregulated by PVC + METH at 5 mg/L ($p < 0.05$) with 14.7-fold, 1.9-fold, and 1.9-fold, respectively. However, the genes *pla2g4ab* and *nfatc2a* showed no effects on the expression levels ($p > 0.05$ vs. the control) in the PVC and PVC + 5 groups. Moreover, significant upregulation of *nfatc4* was found in the PVC and PVC + 5 groups ($p < 0.05$, 4.9-fold and 6.6-fold).

4. Discussion

In this study, the effects on the ecology-associated indicators of zebrafish larvae of the primary MPs (PS and PVC) were estimated. Previous studies have indicated the harmful effects of MPs on fish, including oxidative stress, intestinal damage, aberrant behavior functions, disruption development, and morality in fish [51,52]. Similarly, significantly higher mortality of larvae was found in the exposure groups (PS and PVC) than in the control. Since the survival rate of fish larvae is important for the population's prosperity [52], exposure to PS or PVC might threaten fish flourishment. The histopathological results showed significant depositions of PS and PVC MPs in the digestive tract of fish larvae, which might contribute to MP exposure [53]. Noticeably, the depositions of PVC MPs in the intestinal tracts of zebrafish in the combined exposure groups were much greater than in the PVC single exposure group, indicating that the presence of METH might facilitate the ingestion of PVC MPs by fish. When exposure was combined with METH, the lethal effects of PS + METH at 5 mg/L significantly enhances, suggesting the joint toxicity of the two pollutants. Similarly, Qu et al. (2020) found that METH could be significantly absorbed by PS MPs at 20 mg/L, and the acute toxicity (LC50) to the zooplankton increased upon the co-contamination of PS and METH [41]. However, the addition of METH did not change the lethal effects of PVC, which was consistent with a previous study on PVC and di(2-ethylhexyl) phthalate [54]. The toxic effects of MPs on aquatic vertebrates were mainly associated with oxidative damage [52] and thus induced inflammation in fish [55]. Moreover, METH exposure at low concentrations might induce oxidative stress and morphological developmental abnormalities in zebrafish [56]. Therefore, it should be determined whether the co-exposure of MPs (PS and PVC) and METH enhances the toxicity to zebrafish. The histopathological changes in zebrafish larvae demonstrated that the presence of methamphetamine exacerbates the harmful effects of microplastics on fish, including enlarged trunk development and intestinal hydrops, but there are some differences in the pathological changes observed in the larvae among different microplastics groups, suggesting the different toxicity of varying types of MPs [57].

Since fish are the top predators of aquatic ecosystems, changes in their ecology-associated behaviors are considered to be important biomarkers for evaluating MP toxicity [58]. A previous study has evidenced the neurotoxic effect of PS and PVC MPs on fish, showing inhibition of locomotion [59]. In this study, the same results were found in zebrafish larvae, including the simplified swimming trajectory, extended immobility, and lower mean velocity. Noticeably, exposure to METH showed the same anesthetic effects on zebrafish [56]. The addition of METH significantly alleviated the effects on the immobility duration of larvae induced by PS MPs while having no impact on the PVC group. Meanwhile, the turn angle decreased exposure to PS and PVC MPs, which was then mitigated with combined exposure to METH. The results indicated that the combined exposure with METH might mitigate the adverse effects of MPs on behavioral functions of zebrafish larvae, and showed differences between PS and PVC MPs. From an ecological perspective, fish rely on swimming behavior for predation and avoiding predators in water, and reduced movement activity can diminish their success rates in these activities [60], indirectly leading to increased mortality rates and severe ecological consequences.

At this stage, studies on the toxic mechanisms of MPs on aquatic organisms have found that lipid metabolism and energy metabolism of zebrafish are suppressed by long-term PS MP stress [51]. Exposure to PVC MPs stimulated the lipid metabolic pathway and the phosphoinositide 3 kinase (PI3K)/protein kinase B (Akt) signaling pathway in mice liver [61]. However, the different mechanisms of the PS and PVC MPs, as well as combined exposure with METH, were not determined. Hence, this study for the first time elucidated the underlying mechanisms based on RNA seq. The results revealed that the majority of significantly differentially expressed genes were closely associated with metabolic pathways, such as drug metabolism, glycerolipid metabolism, and steroid synthesis. Therefore, it can be inferred that both single and combined exposures might interfere with metabolic functions in zebrafish larvae. Previous research has indicated that disruptions in fish metabolic functions can impact various aspects of fish health, behavioral traits, and survival rates [62]. The outcomes are consistent with the experimental results of this study, where significant differences were observed in mortality rates, swimming behavior, and histopathological slices of zebrafish larvae exposed solely or jointly to microplastics compared to the control group.

Furthermore, we quantitatively analyzed the expression levels of the genes involved in the MAPK signaling pathway and the C-type lectin signaling pathway of fish in the PS groups. The results indicated that the activation of the MAPK signaling pathway and the C-type lectin pathway in zebrafish larvae might contribute to the adverse effects on their ecology-associated indicators (e.g., mortality and histopathology) induced by a combination of PS and METH. The mitogen-activated protein kinase (MAPK) family consists of a group of serine/threonine kinases that mediate intracellular signal transduction [63]. The MAPK pathways are activated by various extracellular and intracellular stimuli, including peptide growth factors, cytokines, hormones, and various cellular stressors such as oxidative stress and endoplasmic reticulum stress [62]. Hence, the upregulation of the genes (i.e., *srfa*, *elk1*, and *elk2*) of fish observed in the PS + METH group indicated that combined exposure with PS MPs and METH could disrupt the normal physiological activities of cells, thereby influencing the growth and development of the larvae through activating the MAPK signaling pathway [62]. Given that the upregulation of mRNA expression in the genes *jun* and *mapk9*, related to the apoptotic pathway, can promote cell apoptosis [64], the significant increase found in this study (PS and PS + 5) demonstrated that cell apoptosis might contribute to the adverse effects induced by single or joint exposure to PS MPs and METH. C-type lectins are important pattern recognition receptors in the immune system, and the normal expression of genes in the c-type lectin signaling pathway is crucial for the immune system [65]. Genes related to the c-type lectin signaling pathway, *nfatc2a* and *nfatc4*, were significantly upregulated by PS + METH implied that combined exposure induced immune responses of zebrafish larvae. Following the impairment of the normal immune system in juvenile fish, it might lead to increased susceptibility to infections, growth limitations, decreased survival capabilities, and adverse impacts on ecological balance [66].

Meanwhile, we quantitatively analyzed the expression levels of the genes involved into steroid hormone biosynthesis pathways and C-type lectin signaling pathway of fish in PVC groups. The results implied that the combined exposure of PVC and METH might lead to adverse effects on ecology-associated indicators (i.e., mortality and behavioral functions) in zebrafish larvae through the activation of steroid hormone biosynthesis pathways and the C-type lectin signaling pathway. Cytochrome P450 (CYP) enzymes are among the most important metabolic enzymes in the liver. They are involved in the metabolism of endogenous compounds (such as steroids) and exogenous compounds (such as drugs, carcinogens, and toxic substances). Many studies have indicated that the accumulation of these exogenous compounds in the body can induce changes in gene expression and enzyme function in CYP enzymes, leading to a decrease in their pharmacological effects or toxicity [67]. The genes *cyp3c1*, *cyp3c4*, and *cyp19a1a* are involved in the synthesis of these CYP enzymes. Hence, the upregulation of these genes of fish observed in the

PVC + METH group indicated that combined exposure with PVC MPs and METH could affect larvae normal liver metabolism, thereby influencing the growth and development of the larvae through changing the steroid hormone biosynthesis pathways [67]. The gene *nfatc4*, related to the C-type lectin signaling pathway, was significantly upregulated by PVC + METH, implying that combined exposure induced immune responses in zebrafish larvae. When zebrafish larvae experience a compromised immune system, it can result in heightened vulnerability to infections, growth constraints, diminished survival abilities, and detrimental effects on ecological equilibrium [66]. As such, the significant upregulation of gene mRNA expression in the combined exposure group suggested that the combined toxicity of PS + METH or PVC + METH was likely to have a substantial impact on the growth and development of zebrafish larvae, which should be a cause for concern.

5. Conclusions

In summary, this study reveals the combined toxicity of PS and PVC MPs and METH in zebrafish larvae. The results demonstrate increased larval mortality, abnormal behavioral patterns, and histopathological alterations upon exposure to PS and PVC microplastics. The addition of METH might enhance these adverse effects. Meanwhile, transcriptomic data showed that exposure to PS and METH activated MAPK signaling pathways and the C-type lectin signaling pathway, and PVC and METH stimulated steroid hormone synthesis pathways and the C-type lectin signaling pathway, which might then trigger apoptosis and immune responses in fish. This study for the first time elucidated the joint toxicity of primary MPs (PS and PVC) and the illicit drugs (METH) to zebrafish larvae, and the underlying mechanisms were investigated. The results should promote the development of ways of preserving the environment and protecting public health from risks associated with the co-pollution of MPs and pharmaceuticals.

Supplementary Materials: The following supporting information can be downloaded at: https://www.mdpi.com/article/10.3390/toxics12010064/s1, Figure S1: title; Table S1: Primer sequences for the genes tested in the present study; Figure S1: Purity of PS and PVC microplastics identified by Fourier transform infrared spectroscopy; Figure S2: Electron microscope scans and particle size of PS and PVC microplastics, Monomer of PS (a–c) and PVC (d–f) MPs showed as smooth and intact sphere; Figure S3: The correlation networks of PS or PS and METH based on the DEGs and the enrichment pathways; Figure S4: The correlation networks of PVC or PVC and METH based on the DEGs and the enrichment pathways.

Author Contributions: H.W., Z.W.: Investigation, Writing—Original Draft, Writing—Review and Editing. J.X., Y.Y., and W.Z.: Formal Analysis, Data Curation. J.L.: Conceptualization, Supervision, Resources. All authors have read and agreed to the published version of the manuscript.

Funding: This study received financial support from the Natural Science Foundation of Jiangsu Province (BK20210376) and Ningbo Public Welfare Technology Plan Project (2022S118).

Institutional Review Board Statement: All animal studies were performed in accordance with the Guidelines for Animal Experiments of Hohai University, which meet the ethical guidelines for experimental animals in China.

Informed Consent Statement: Not applicable.

Data Availability Statement: Data are contained within the article.

Conflicts of Interest: The manuscript has been read and approved by all authors, and no conflicts of interest exist in the submission of this manuscript. Neither the entire paper nor any part of its content has been published or has been accepted elsewhere. It is not being submitted to any other journal.

References

1. Barboza, L.G.A.; Dick Vethaak, A.; Lavorante, B.R.B.O.; Lundebye, A.-K.; Guilhermino, L. Marine microplastic debris: An emerging issue for food security, food safety and human health. *Mar. Pollut. Bull.* **2018**, *133*, 336–348. [CrossRef] [PubMed]
2. Cole, M.; Lindeque, P.; Halsband, C.; Galloway, T.S. Microplastics as contaminants in the marine environment: A review. *Mar. Pollut. Bull.* **2011**, *62*, 2588–2597. [CrossRef] [PubMed]

3. da Costa, J.P.; Santos, P.S.M.; Duarte, A.C.; Rocha-Santos, T. (Nano)plastics in the environment—Sources, fates and effects. *Sci. Total Environ.* **2016**, *566–567*, 15–26. [CrossRef] [PubMed]
4. Gonçalves, C.; Martins, M.; Sobral, P.; Costa, P.M.; Costa, M.H. An assessment of the ability to ingest and excrete microplastics by filter-feeders: A case study with the Mediterranean mussel. *Environ. Pollut.* **2019**, *245*, 600–606. [CrossRef]
5. Thompson, R.C.; Olsen, Y.; Mitchell, R.P.; Davis, A.; Rowland, S.J.; John, A.W.G.; McGonigle, D.; Russell, A.E. Lost at Sea: Where Is All the Plastic? *Science* **2004**, *304*, 838. [CrossRef] [PubMed]
6. Aydın, İ.; Terzi, Y.; Gündoğdu, S.; Aytan, Ü.; Öztürk, R.Ç.; Atamanalp, M.; Alak, G.; Sivri, N.; Akarsu, C.; Atıcı, A.A. Microplastic Pollution in Turkish Aquatic Ecosystems: Sources, Characteristics, Implications, and Mitigation Strategies. *Turk. J. Fish. Aquat. Sci.* **2023**, *23*, 24773.
7. Lechner, A.; Keckeis, H.; Lumesberger-Loisl, F.; Zens, B.; Krusch, R.; Tritthart, M.; Glas, M.; Schludermann, E. The Danube so colourful: A potpourri of plastic litter outnumbers fish larvae in Europe's second largest river. *Environ. Pollut.* **2014**, *188*, 177–181. [CrossRef]
8. Carlin, J.; Craig, C.; Little, S.; Donnelly, M.; Fox, D.; Zhai, L.; Walters, L. Microplastic accumulation in the gastrointestinal tracts in birds of prey in central Florida, USA. *Environ. Pollut.* **2020**, *264*, 114633. [CrossRef]
9. Mallik, A.; Xavier, K.A.M.; Naidu, B.C.; Nayak, B.B. Ecotoxicological and physiological risks of microplastics on fish and their possible mitigation measures. *Sci. Total Environ.* **2021**, *779*, 146433. [CrossRef]
10. Gambardella, C.; Morgana, S.; Ferrando, S.; Bramini, M.; Piazza, V.; Costa, E.; Garaventa, F.; Faimali, M. Effects of polystyrene microbeads in marine planktonic crustaceans. *Ecotoxicol. Environ. Saf.* **2017**, *145*, 250–257. [CrossRef]
11. Khalid, N.; Aqeel, M.; Noman, A.; Hashem, M.; Mostafa, Y.S.; Alhaithloul, H.A.S.; Alghanem, S.M. Linking effects of microplastics to ecological impacts in marine environments. *Chemosphere* **2021**, *264*, 128541. [CrossRef] [PubMed]
12. Atamanalp, M.; Kırıcı, M.; Köktürk, M.; Kırıcı, M.; Kocaman, E.M.; Ucar, A.; Parlak, V.; Özcan, S.; Yanık, T.; Alak, G. Polyethylene exposure in rainbow trout; suppresses growth and may act as a promoting agent in tissue-based oxidative response, DNA damage and apoptosis. *Process Saf. Environ. Prot.* **2023**, *174*, 960–970. [CrossRef]
13. Köktürk, M.; Özgeriş, F.B.; Atamanalp, M.; Ucar, A.; Özdemir, S.; Parlak, V.; Duyar, H.A.; Alak, G. Microplastic-induced oxidative stress response in turbot and potential intake by humans. *Drug Chem. Toxicol.* **2023**, 1–10. [CrossRef] [PubMed]
14. Alak, G.; Köktürk, M.; Atamanalp, M.; Kocaman, E.M.; Ucar, A.; Esenbuğa, N.; Özcan, S.; Parlak, V. Microplastic Abundance in Rainbow Trout Life Cycle: Step by Step. *Sustainability* **2023**, *15*, 14255. [CrossRef]
15. Alimba, C.G.; Faggio, C. Microplastics in the marine environment: Current trends in environmental pollution and mechanisms of toxicological profile. *Environ. Toxicol. Pharmacol.* **2019**, *68*, 61–74. [CrossRef]
16. Fernández-González, V.; Andrade-Garda, J.M.; López-Mahía, P.; Muniategui-Lorenzo, S. Misidentification of PVC microplastics in marine environmental samples. *TrAC Trends Anal. Chem.* **2022**, *153*, 116649. [CrossRef]
17. Lithner, D.; Damberg, J.; Dave, G.; Larsson, Å. Leachates from plastic consumer products—Screening for toxicity with *Daphnia magna*. *Chemosphere* **2009**, *74*, 1195–1200. [CrossRef] [PubMed]
18. Iheanacho, S.C.; Odo, G.E. Neurotoxicity, oxidative stress biomarkers and haematological responses in African catfish (*Clarias gariepinus*) exposed to polyvinyl chloride microparticles. *Comp. Biochem. Physiol. Part C Toxicol. Pharmacol.* **2020**, *232*, 108741. [CrossRef]
19. Sun, S.; Jin, Y.; Luo, P.; Shi, X. Polystyrene microplastics induced male reproductive toxicity and transgenerational effects in freshwater prawn. *Sci. Total Environ.* **2022**, *842*, 156820. [CrossRef]
20. Della Torre, C.; Bergami, E.; Salvati, A.; Faleri, C.; Cirino, P.; Dawson, K.A.; Corsi, I. Accumulation and Embryotoxicity of Polystyrene Nanoparticles at Early Stage of Development of Sea Urchin Embryos Paracentrotus lividus. *Environ. Sci. Technol.* **2014**, *48*, 12302–12311. [CrossRef]
21. Hollerova, A.; Hodkovicova, N.; Blahova, J.; Faldyna, M.; Franc, A.; Pavlokova, S.; Tichy, F.; Postulkova, E.; Mares, J.; Medkova, D. Polystyrene microparticles can affect the health status of freshwater fish–Threat of oral microplastics intake. *Sci. Total Environ.* **2023**, *858*, 159976. [CrossRef] [PubMed]
22. Wang, Z.; Wang, Y.; Qin, S.; Yang, Z.; Sun, Y. Polystyrene microplastics weaken the predator-induced defenses of *Daphnia magna*: Evidences from the changes in morphology and behavior. *Environ. Pollut.* **2023**, *316*, 120657. [CrossRef] [PubMed]
23. Liu, G.; Zhu, Z.; Yang, Y.; Sun, Y.; Yu, F.; Ma, J. Sorption behavior and mechanism of hydrophilic organic chemicals to virgin and aged microplastics in freshwater and seawater. *Environ. Pollut.* **2019**, *246*, 26–33. [CrossRef]
24. Yang, Y.; Liu, G.; Song, W.; Ye, C.; Lin, H.; Li, Z.; Liu, W. Plastics in the marine environment are reservoirs for antibiotic and metal resistance genes. *Environ. Int.* **2019**, *123*, 79–86. [CrossRef] [PubMed]
25. Zhu, Z.-L.; Wang, S.-V.; Zhao, F.-F.; Wang, S.-G.; Liu, F.-F.; Liu, G.-Z. Joint toxicity of microplastics with triclosan to marine microalgae *Skeletonema costatum*. *Environ. Pollut.* **2019**, *246*, 509–517. [CrossRef]
26. Prata, J.C.; Lavorante, B.R.B.O.; Montenegro, M.d.C.B.S.M.; Guilhermino, L. Influence of microplastics on the toxicity of the pharmaceuticals procainamide and doxycycline on the marine microalgae *Tetraselmis chuii*. *Aquat. Toxicol.* **2018**, *197*, 143–152. [CrossRef]
27. Wang, Y.; Zhao, Y.; Liang, H.; Ma, C.; Cui, N.; Cao, H.; Wei, W.; Liu, Y. Single and combined effects of polyethylene microplastics and acetochlor on accumulation and intestinal toxicity of zebrafish (*Danio rerio*). *Environ. Pollut.* **2023**, *333*, 122089. [CrossRef]

28. Byeon, E.; Jeong, H.; Lee, Y.-J.; Cho, Y.; Lee, K.-W.; Lee, E.; Jeong, C.-B.; Lee, J.-S.; Kang, H.-M. Effects of microplastics and phenanthrene on gut microbiome and metabolome alterations in the marine medaka *Oryzias melastigma*. *J. Hazard. Mater.* **2024**, *461*, 132620. [CrossRef]
29. O'Malley, K.Y.; Hart, C.L.; Casey, S.; Downey, L.A. Methamphetamine, amphetamine, and aggression in humans: A systematic review of drug administration studies. *Neurosci. Biobehav. Rev.* **2022**, *141*, 104805. [CrossRef]
30. Heal, D.J.; Smith, S.L.; Gosden, J.; Nutt, D.J. Amphetamine, past and present—A pharmacological and clinical perspective. *J. Psychopharmacol.* **2013**, *27*, 479–496. [CrossRef]
31. Wang, Z.; Han, S.; Cai, M.; Du, P.; Zhang, Z.; Li, X. Environmental behavior of methamphetamine and ketamine in aquatic ecosystem: Degradation, bioaccumulation, distribution, and associated shift in toxicity and bacterial community. *Water Res.* **2020**, *174*, 115585. [CrossRef] [PubMed]
32. Cruickshank, C.C.; Dyer, K.R. A review of the clinical pharmacology of methamphetamine. *Addiction* **2009**, *104*, 1085–1099. [CrossRef] [PubMed]
33. Panenka, W.J.; Procyshyn, R.M.; Lecomte, T.; MacEwan, G.W.; Flynn, S.W.; Honer, W.G.; Barr, A.M. Methamphetamine use: A comprehensive review of molecular, preclinical and clinical findings. *Drug Alcohol Depend.* **2013**, *129*, 167–179. [CrossRef] [PubMed]
34. Wongprayoon, P.; Govitrapong, P. Melatonin Protects SH-SY5Y Neuronal Cells Against Methamphetamine-Induced Endoplasmic Reticulum Stress and Apoptotic Cell Death. *Neurotox. Res.* **2016**, *31*, 1–10. [CrossRef] [PubMed]
35. Foroughi, K.; Khaksari, M.; Rahmati, M.; Bitaraf, F.S.; Shayannia, A. Apelin-13 Protects PC12 Cells Against Methamphetamine-Induced Oxidative Stress, Autophagy and Apoptosis. *Neurochem. Res.* **2019**, *44*, 2103–2112. [CrossRef] [PubMed]
36. Li, J.; Zhu, L.; Su, H.; Liu, D.; Yan, Z.; Ni, T.; Wei, H.; Goh, E.L.K.; Chen, T. Regulation of miR-128 in the nucleus accumbens affects methamphetamine-induced behavioral sensitization by modulating proteins involved in neuroplasticity. *Addict. Biol.* **2020**, *26*, e12881. [CrossRef] [PubMed]
37. Toborek, M.; Seelbach, M.J.; Rashid, C.S.; András, I.E.; Chen, L.; Park, M.; Esser, K.A. Voluntary exercise protects against methamphetamine-induced oxidative stress in brain microvasculature and disruption of the blood–brain barrier. *Mol. Neurodegener.* **2013**, *8*, 22. [CrossRef]
38. Shaerzadeh, F.; Streit, W.J.; Heysieattalab, S.; Khoshbouei, H. Methamphetamine neurotoxicity, microglia, and neuroinflammation. *J. Neuroinflamm.* **2018**, *15*, 341. [CrossRef]
39. Sancho Santos, M.E.; Horký, P.; Grabicová, K.; Steinbach, C.; Hubená, P.; Šálková, E.; Slavík, O.; Grabic, R.; Randák, T. From metabolism to behaviour—Multilevel effects of environmental methamphetamine concentrations on fish. *Sci. Total Environ.* **2023**, *878*, 163167. [CrossRef]
40. Wang, Z.; Xu, Z.; Li, X. Impacts of methamphetamine and ketamine on C.elegans's physiological functions at environmentally relevant concentrations and eco-risk assessment in surface waters. *J. Hazard. Mater.* **2019**, *363*, 268–276. [CrossRef]
41. Qu, H.; Ma, R.; Barrett, H.; Wang, B.; Han, J.; Wang, F.; Chen, P.; Wang, W.; Peng, G.; Yu, G. How microplastics affect chiral illicit drug methamphetamine in aquatic food chain? From green alga (*Chlorella pyrenoidosa*) to freshwater snail (*Cipangopaludian cathayensis*). *Environ. Int.* **2020**, *136*, 105480. [CrossRef] [PubMed]
42. Horzmann, K.A.; Freeman, J.L. Making Waves: New Developments in Toxicology With the Zebrafish. *Toxicol. Sci.* **2018**, *163*, 5–12. [CrossRef] [PubMed]
43. Mu, X.; Qi, S.; Liu, J.; Yuan, L.; Huang, Y.; Xue, J.; Qian, L.; Wang, C.; Li, Y. Toxicity and behavioral response of zebrafish exposed to combined microplastic and bisphenol analogues. *Environ. Chem. Lett.* **2022**, *20*, 41–48. [CrossRef]
44. Ding, N.; Jiang, L.; Wang, X.; Wang, C.; Geng, Y.; Zhang, J.; Sun, Y.; Zhang, Y.; Yuan, Q.; Liu, H. Polyethylene microplastic exposure and concurrent effect with Aeromonas hydrophila infection on zebrafish. *Environ. Sci. Pollut. Res.* **2022**, *29*, 63964–63972. [CrossRef] [PubMed]
45. Hamed, M.; Soliman, H.A.; Osman, A.G.; Sayed, A.E.-D.H. Assessment the effect of exposure to microplastics in Nile Tilapia (*Oreochromis niloticus*) early juvenile: I. blood biomarkers. *Chemosphere* **2019**, *228*, 345–350. [CrossRef]
46. Mi, G.; Gao, Y.; Yan, H.; Jin, X.; Ye, E.; Liu, S.; Gong, Z.; Yang, H.; Yang, Z. l-Scoulerine attenuates behavioural changes induced by methamphetamine in zebrafish and mice. *Behav. Brain Res.* **2016**, *298*, 97–104. [CrossRef] [PubMed]
47. Liao, P.-H.; Hwang, C.-C.; Chen, T.-H.; Chen, P.-J. Developmental exposures to waterborne abused drugs alter physiological function and larval locomotion in early life stages of medaka fish. *Aquat. Toxicol.* **2015**, *165*, 84–92. [CrossRef]
48. Maeda, H.; Hasumi, A.; Yoshida, K.-i. Caffeine-induced bradycardia, death, and anxiety-like behavior in zebrafish larvae. *Forensic Toxicol.* **2021**, *39*, 427–436. [CrossRef]
49. Zakhary, S.M.; Ayubcha, D.; Ansari, F.; Kamran, K.; Karim, M.; Leheste, J.R.; Horowitz, J.M.; Torres, G. A behavioral and molecular analysis of ketamine in zebrafish. *Synapse* **2010**, *65*, 160–167. [CrossRef]
50. Riehl, R.; Kyzar, E.; Allain, A.; Green, J.; Hook, M.; Monnig, L.; Rhymes, K.; Roth, A.; Pham, M.; Razavi, R.; et al. Behavioral and physiological effects of acute ketamine exposure in adult zebrafish. *Neurotoxicol. Teratol.* **2011**, *33*, 658–667. [CrossRef]
51. Lu, Y.; Zhang, Y.; Deng, Y.; Jiang, W.; Zhao, Y.; Geng, J.; Ding, L.; Ren, H. Uptake and Accumulation of Polystyrene Microplastics in Zebrafish (*Danio rerio*) and Toxic Effects in Liver. *Environ. Sci. Technol.* **2016**, *50*, 4054–4060. [CrossRef] [PubMed]
52. De Marco, G.; Conti, G.O.; Giannetto, A.; Cappello, T.; Galati, M.; Iaria, C.; Pulvirenti, E.; Capparucci, F.; Mauceri, A.; Ferrante, M.; et al. Embryotoxicity of polystyrene microplastics in zebrafish *Danio rerio*. *Environ. Res.* **2022**, *208*, 112552. [CrossRef]

53. Hamed, M.; Soliman, H.A.M.; Badrey, A.E.A.; Osman, A.G.M. Microplastics induced histopathological lesions in some tissues of tilapia (*Oreochromis niloticus*) early juveniles. *Tissue Cell* **2021**, *71*, 101512. [CrossRef] [PubMed]
54. Wang, H.; Wang, Y.; Wang, Q.; Lv, M.; Zhao, X.; Ji, Y.; Han, X.; Wang, X.; Chen, L. The combined toxic effects of polyvinyl chloride microplastics and di(2-ethylhexyl) phthalate on the juvenile zebrafish (*Danio rerio*). *J. Hazard. Mater.* **2022**, *440*, 129711. [CrossRef] [PubMed]
55. Pirsaheb, M.; Hossini, H.; Makhdoumi, P. Review of microplastic occurrence and toxicological effects in marine environment: Experimental evidence of inflammation. *Process Saf. Environ. Prot.* **2020**, *142*, 1–14. [CrossRef]
56. Yin, X.; Deng, Y.; Guo, C.; Ding, C.; Xu, J.; Wu, F. Behavioral Changes and Metabolic Responses of Adult Zebrafish (*Danio Rerio*) Exposed to Methamphetamine. *ACS EST Water* **2023**, *3*, 2551–2559. [CrossRef]
57. Wang, C.; Zhao, J.; Xing, B. Environmental source, fate, and toxicity of microplastics. *J. Hazard. Mater.* **2021**, *407*, 124357. [CrossRef]
58. Jambeck, J.R.; Geyer, R.; Wilcox, C.; Siegler, T.R.; Perryman, M.; Andrady, A.; Narayan, R.; Law, K.L. Plastic waste inputs from land into the ocean. *Science* **2015**, *347*, 768–771. [CrossRef]
59. Kim, J.-H.; Yu, Y.-B.; Choi, J.-H. Toxic effects on bioaccumulation, hematological parameters, oxidative stress, immune responses and neurotoxicity in fish exposed to microplastics: A review. *J. Hazard. Mater.* **2021**, *413*, 125423. [CrossRef]
60. Qiang, L.; Cheng, J. Exposure to microplastics decreases swimming competence in larval zebrafish (*Danio rerio*). *Ecotoxicol. Environ. Saf.* **2019**, *176*, 226–233. [CrossRef]
61. Chen, X.; Zhuang, J.; Chen, Q.; Xu, L.; Yue, X.; Qiao, D. Chronic exposure to polyvinyl chloride microplastics induces liver injury and gut microbiota dysbiosis based on the integration of liver transcriptome profiles and full-length 16S rRNA sequencing data. *Sci. Total Environ.* **2022**, *839*, 155984. [CrossRef] [PubMed]
62. Hu, Q.; Wang, H.; He, C.; Jin, Y.; Fu, Z. Polystyrene nanoparticles trigger the activation of p38 MAPK and apoptosis via inducing oxidative stress in zebrafish and macrophage cells. *Environ. Pollut.* **2021**, *269*, 116075. [CrossRef] [PubMed]
63. Coulthard, L.R.; White, D.E.; Jones, D.L.; McDermott, M.F.; Burchill, S.A. p38MAPK: Stress responses from molecular mechanisms to therapeutics. *Trends Mol. Med.* **2009**, *15*, 369–379. [CrossRef] [PubMed]
64. Li, Y.; Wang, R.; Li, Y.; Sun, G.; Mo, H. Protective effects of tree peony seed protein hydrolysate on Cd-induced oxidative damage, inflammation and apoptosis in zebrafish embryos. *Fish Shellfish Immunol.* **2022**, *126*, 292–302. [CrossRef]
65. Wang, L.; Zhang, J.; Zhao, X.; Pei, C.; Li, L.; Kong, X. Molecular characterization and biological effect of a C-type lectin receptor in Qihe crucian carp, Carassius auratus. *Dev. Comp. Immunol.* **2021**, *121*, 104081. [CrossRef]
66. Aksakal, F.I.; Ciltas, A. Impact of copper oxide nanoparticles (CuO NPs) exposure on embryo development and expression of genes related to the innate immune system of zebrafish (*Danio rerio*). *Comp. Biochem. Physiol. Part C Toxicol. Pharmacol.* **2019**, *223*, 78–87. [CrossRef]
67. Li, Y.; Liang, H.; Ren, B.; Zhao, T.; Chen, H.; Zhao, Y.; Liang, H. Enantioselective toxic effects of mefentrifluconazole in the liver of adult zebrafish (*Danio rerio*) based on transcription level and metabolomic profile. *Toxicology* **2022**, *467*, 153095. [CrossRef]

toxics MDPI

Article

Methamphetamine Shows Different Joint Toxicity for Different Types of Microplastics on Zebrafish Larvae by Mediating Oxidative Stress

Jindong Xu [1], Wenqi Yang [1], Dongyi Wang [1], Zhenglu Wang [2], Chuang Liu [1] and Jiana Li [3,*]

1 College of Oceanography, Hohai University, Nanjing 210098, China; aukuma20@gmail.com (J.X.); eliomurphy8@gmail.com (W.Y.); wdynn@foxmail.com (D.W.); huangliu2020@hhu.edu.cn (C.L.)
2 West China School of Public Health and West China Fourth Hospital, Sichuan University, Chengdu 610041, China; m18018567955@163.com
3 Ningbo Academy of Ecological and Environmental Sciences, Ningbo 315000, China
* Correspondence: stlijiana@163.com; Tel.: +86-185-1258-1998

Abstract: The coexistence of polystyrene (PS) and polypropylene (PVC) microplastics (MPs) and methamphetamine (METH) in aquatic systems is evident. However, the joint toxicity is unclear. Here, zebrafish larvae were exposed to single PS and PVC MPs (20 mg L^{-1}) and combined with METH (250 and 500 μg L^{-1}) for 10 days. The results indicated that acute exposure to PS and PVC MPs induced lethal effects on zebrafish larvae (10–20%). Treatment with MPs markedly suppressed the locomotion of zebrafish, showing as the lengthy immobility (51–74%) and lower velocity (0.09–0.55 cm s^{-1}) compared with the control (1.07 cm s^{-1}). Meanwhile, histopathological analysis revealed pronounced depositions of MPs particles in fish's intestinal tract, triggering inflammatory responses (histological scores: 1.6–2.0). In the coexposure groups, obviously inflammatory responses were found. Furthermore, the up-regulations of the genes involved in the oxidative kinase gene and inflammation related genes implied that oxidative stress triggered by MPs on zebrafish larvae might be responsible for the mortality and locomotion retardant. The antagonistic and stimulatory effects of METH on the expression changes of genes found in PVC and PS groups implied the contrary combined toxicity of PS/PVC MPs and METH. This study for the first time estimated the different toxicity of PS and PVC MPs on fish and the joint effects with METH at high environmental levels. The results suggested PS showed stronger toxicity than PVC for fish larvae. The addition of METH stimulated the effects of PS but antagonized the effects of PVC, promoting control strategy development on MPs and METH in aquatic environments.

Keywords: microplastics; methamphetamine; joint toxicity; behavioral functions; oxidative stress

1. Introduction

In conjunction with the growth of the plastic manufacturing industry, microplastics (MPs)—small particles derived from plastic waste—have emerged as environmental pollutants in aquatic systems [1]. Statistically, the total yield of plastic production in 2018 was 359 million tons worldwide, with 5–10% of that drained into the ocean [2]. Over time, plastic waste breaks down into smaller particles through the action of microorganisms, light, and hydraulic erosion, with particles smaller than 5 mm in diameter classified as MPs [3]. According to the monitoring results, the concentrations of MPs in the Yangtze Estuary and Taihu Lake were 10,200 items m^{-3} and 25,800 items m^{-3}, respectively [4,5]. In the Rhine River, 892,777 MP items per square kilometer were detected [6]. However, it is crucial to emphasize that these measurements mainly pertain to MPs with larger particle sizes. A survey revealed that, for MPs with smaller sizes, 1–10 μm MPs constitute approximately 80% of the pollution found in water bodies [7]. Among them, polystyrene (PS), polyethylene (PE), and polypropylene (PP) are the predominant MPs and have been frequently detected in marine animals, such as mussels and fishes [8]. When ingested by organisms through feeding, MPs can induce a range of adverse effects, including oxidative stress, behavioral dysfunctions, immunological responses, and morphological damage [9–12]. Even in the absence of ingestion, MPs may still pose a threat to organisms. Previous studies have shown that MPs can adsorb onto the envelope of zebrafish embryos after short-term exposure, leading to aberrant behaviors in the larvae [10]. Hence, the threat of MPs to the environment should be considered. Furthermore, the outcomes of MPs may vary depending on their characteristics, such as polymer types, additives, and surface properties, leading to heterogeneity in their effects [13,14]. The underlying mechanisms of such opposite effects are still unclear.

In China, PS is a major contributor to MPs pollution in water, with an annual production of up to 1.8 million tons [15]. Indeed, previous research has provided evidence of the toxicity of larger-sized PS, typically ranging from 75 to 200 μm, on zooplankton organisms [16]. It is well known that the average size of PS MPs in surface water falls within the range of 10–20 μm [17]. However, there is a lack of comprehensive assessment regarding the potential impairments specifically posed by MPs of smaller sizes. The existing studies indicate that surface modification and pH can influence the aggregation of nano-sized PS MPs in aquatic environments. Furthermore, it has been confirmed that these factors can cause damage to the digestive tract of large crayfish [18]. However, at a concentration of 1 mg/L, no significant toxicity was observed in zebrafish [19]. Additionally, it is important to note that PVC is another commonly used plastic that may have toxic effects on organisms due to its functional groups and additives [20]. Previous studies have shown that PVC can inhibit the increase in body weight and length of *Cyprinus carpio* var. larvae and cause oxidative stress damage and hepatocyte vacuolation [21]. Another study showed that PVC can inhibit the expression of genes related to heart development in zebrafish larvae, resulting in the weakening of heart function [22]. Nevertheless, the adverse effects on fishes lack more evidence.

Moreover, MPs feature the potential to act as chemical vectors for some organic chemicals in aquatic environments due to their large specific surface area and high absorbability [23]. This finding derives a new scenario for coexposure to MPs and organic contaminants, which may lead to joint effects on organisms. When ingested by organisms, coexposure to MPs and other environmental contaminants could enhance the bioavailability and adverse effects on biota compared to monomer MPs exposure, including increased accumulation of organic pollutants, abnormal glycolipid metabolism, hepatotoxicity, and endocrine disruption [24–26]. However, the combined toxicity of MPs and organic pollutants is extremely complex. For example, the presence of PS MPs significantly decreased the harm to mussels caused by PAHs [27] but increased the toxicity of triclosan to marine copepods [28]. Meanwhile, some studies have shown that the toxicity of PVC and DEHP can be antagonized, and it is believed that PVC can reduce its toxicity by increasing the adsorption of DEHP [29]. The underlying mechanisms of such opposite effects are still unclear. Nevertheless, it can be ascertained that variations in the adsorption capacity of different types of MPs and organic pollutants exist [30,31]. The adsorption relationship between MPs and organic pollutants can also be influenced by factors such as changes in pH, the presence of cations, and the aging degree of water [30,31]. These factors are considered to be contributors to the observed differences in the joint effects of MPs and organic pollutants. Further research is required to fully elucidate the intricate mechanisms underlying these phenomena [32,33].

As emerging organic pollutants, illicit drugs are of great concern due to their ubiquity, teratogenic potency, and neurotoxicity [22,34]. Methamphetamine (METH) is the primary illicit drug widely detected in the surface water of China, with the highest concentrations up to 405 ng L^{-1} [35]. In addition, METH is also widely detected worldwide. In the Nepean River of New South Wales in Australia, the detected concentration is up to 25.25 μg L^{-1} [36], and in the Llobregat basin in Spain, the detected concentration is 50 ng L^{-1} [37]. Previous studies found that exposure to METH at low environmental levels could induce obvious malformation of zebrafish larvae [12,38]. Meanwhile, another article shows that exposure to METH can lead to behavioral abnormalities and certain reproductive toxicity in guppies [22]. In addition, studies have shown that METH at environmental levels can lead to wild brown trout addiction and change habitat preferences, which may have adverse consequences at the population level [39]. The persistent occurrence of METH in water might result from its interaction with MPs. Qu et al. reported that METH is adsorbed by PS MPs and promotes their bioavailability [40], suggesting that it is warranted to estimate the synergistic toxicity of MPs and METH.

Therefore, the toxicity of individual PS and PVC MPs and the joint effects of MPs and METH on zebrafish larvae were investigated in this work. The aims were to (1) elucidate the adverse effects on larval developmental processes, (2) reveal the impacts on the locomotor activity of fishes, (3) demonstrate the histopathological alterations of the larvae, and (4) elucidate the molecular mechanisms. Through comprehensive analysis and exploration of the aforementioned aspects, it is possible to investigate whether the presence of MPs enhances or attenuates the toxicity of METH.

2. Materials and Methods

2.1. Animals and Experimental Design

Zebrafish (Danio rerio) embryos were obtained from Shanghai Fei Xi Biotechnology Co. (Shanghai, China) and incubated at the Department of Marine Biology, Hohai University, according to the animal research protocol approved by the Institutional Animal Care and Use Committee. Embryos were cultivated in embryo-rearing media (ERM; containing 1 g NaCl, 0.03 g KCl, 0.04 g $CaCl_2 \cdot 2H_2O$, and 0.163 g $MgSO_4 \cdot 7H_2O$ in 1 L DD water, adjusted to pH 7.2) until hatching [41]. After hatching, the larvae were incubated in cultivation water with the following parameters: dissolved oxygen > 7.0 mg L^{-1}; pH 7.0–7.4; and no detectable residual chlorine, ammonia, or nitrite. Water quality and environmental conditions were rigorously maintained, with a 14:10 light:dark photoperiod at 26 ± 1 °C. During cultivation, the parental zebrafish were fed twice daily with frozen brine shrimp (Artemia nauplii) from a fresh source. Each breeding group consisted of 6 male zebrafish and 3 female zebrafish. Spawning behavior was induced using light after a period of darkness, and, after 4 h of fertilization, the fertilized eggs were collected.

The standards PS and PVC MPs (average diameter: 10 μm) were purchased from Dongguan Junxin Plastics Co. Ltd., (Dongguan, China). The purity and characteristics were identified by a Thermo Scientific Nicolet iN10 infrared spectrometer (Thermo Scientific, Waltham, MA, USA). The distributions of particle size were determined using a Malvern mastersizer 2000 laser particle sizer (Malvern Ltd., Malvern, UK). Hydrochloride salts of METH were purchased from Cerilliant (Cerilliant, 1.0 mg mL^{-1} in methanol). The stock solution was diluted using ERM to prepare the working solution of METH for use.

Embryos at the early blastula stage (24 h post fertilization) were randomly placed in a fresh 60-mm crystallizing dish (20 embryos per dish) spiked with MPs or METH. The exposure lasted for 10 d [42]. Seven groups were set: CK group (ERM blank control); PS group (PS MPs at 20 mg L^{-1}); PVC group (PVC MPs at 20 mg L^{-1}); PS250 group (PS MPs 20 mg L^{-1} + METH 250 μg L^{-1}); PS500 group PS MPs (20 mg L^{-1} + METH 500 μg L^{-1}); PVC250 group (PVC MPs 20 mg L^{-1} + METH 250 μg L^{-1}); PVC500 group (PVC MPs 20 mg L^{-1} + METH 500 μg L^{-1}). Each group was replicated with three dishes, and 100% of the exposure solution was replaced every 24 h by transferring the animals into a new dish spiked with fresh exposure solution. All animal studies were performed in accordance with the Guidelines for Animal Experiments of Hohai University, which meet the ethical guidelines for experimental animals in China. To ensure that the dose of MPs was constant, the exposure solution was prepared by weighing in precision using an ultramicroanalytical balance (accurate to 0.1 μg) and used immediately afterward. The levels of METH in the dosing solution were confirmed using HPLC-MS/MS, as previously described [43]. The actual concentrations were 99.3–100.9% of the nominal values (Table S1), and the details are shown in the Supporting Information. During exposure, the mortality and morphology of the embryos were recorded daily using a stereomicroscope (Shang Guang, Suzhou Instrument Co., Ltd., Suzhou, China). At the end of the experiment, embryos were rinsed using fresh ERM three times, and the associated biomarkers were analyzed immediately. The timeline of this work is shown in Figure 1a.

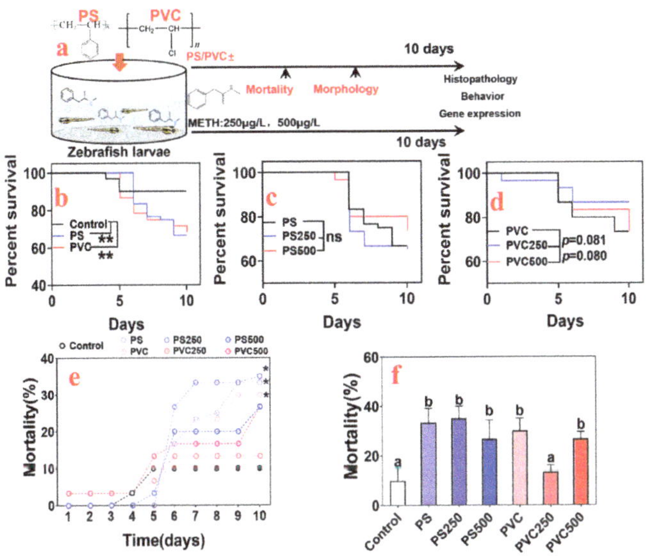

Figure 1. (**a**) The timeline of the experiments: zebrafish larvae (24 h post-fertilization) were exposed to PS MPs, PVC MPs, PS MPs + METH, and PVC MPs + METH for 10 days; during the exposure, the mortality and morphological abnormality of fish were recorded; at the end of the exposure, the histopathological, behavioral, and genetic biomarkers of fish were analyzed. (**b–d**) The survival curves of animals during the exposure in different groups. (**e**) The trajectory curves of mortality at different time points in different groups. (**f**) The terminal mortality of fish in different groups. * $p < 0.05$, ns > 0.05 vs. the control. Distinct letters denote statistically significant differences among the respective groupings.

2.2. *High-Performance Liquid Chromatography Tandem Mass Spectrometry (HPLC-MS/MS) Detection Method*

The 1 mL of medium spiked with METH was pipetted and then filtrated through 0.22 μm cellulose acetate membrane filters. The filtrate (300 μL) was added with an internal standard of METH-d8 with the ultimate dose at 200 ng. Subsequently, chemicals in the mixture were quantified using a UFLCXR-LC system (Shimadzu, Japan) with a Phenomenex Gemini C18 column (100 mm × 2 mm, 3 μm) and an injection volume of 5 μL. The mobile phase was composed of 30 mM ammonium formate in ultrapure water with pH adjusted to 3 using formic acid (98% in water) (A) and 0.1% formic acid in AcN (B). The flow rate of the mobile phase was controlled at 0.3 mL min^{-1}. The elution gradient was as follows: 0–0.1 min: 5% B; 0.1–3.0 min: 30% B; 3.0–5.0 min: 80% B; 5.0–7.5 min: 90% B; 7.5–7.6 min: 5% B; 7.6–13.0 min: 5% B. Concentrations were determined using an API 4000 triple quadrupole mass spectrometer (AB SCIEX, MA, USA) equipped with an electrospray interface operating in positive ionization mode. The quantification of MS system was operated in multiple reaction monitoring (MRM) mode.

2.3. *Behavioral Function Measurement*

At the end of the exposure, eleven fish were collected randomly from each group and subsequently placed into a 6-well plate (one fish per well) spiked with 2.5 mL fresh ERM without MPs and METH. The test was conducted in the light photoperiod after acclimation for 20 min. The locomotion video of each fish was recorded using a CCD camera (Mingchuangda Co. Ltd., Shenzhen, China) for 5 min. After analysis of the baseline, 2 min of the video in which the fishes were at steady state was used to analyze the behavioral parameters (i.e., immobility duration, mean velocity, and turn angle) using ImageJ 1.53u software (National Institutes of Health, Bethesda, MD, USA) with the built-in analysis tool [44]. The immobility duration (s min^{-1}) was defined as the proportion of the duration that the fish kept still of the total measured time. The mean velocity (cm s^{-1}) was the distance traveled per unit time. The turn angle (degree) was the change in movement direction (clockwise or anti-clockwise).

2.4. Histological Examination

After exposure, five zebrafish larvae were randomly collected from each group, euthanized by rapid cooling, fixed in 4% paraformaldehyde (PFA) for more than 24 h, dehydrated in a graded ethanol series, and embedded in paraffin. Sections of the larvae were cut at 8 μm and stained with hematoxylin and eosin (H&E). The obtained slices were examined by a light microscope equipped with an ocular micrometer for alterations in histopathology. The rank evaluations on the histopathological changes of zebrafish were determined using the histopathological assessment tools provided previously [45]: Hemorrhage, hyperemia, and inflammation infiltration scored as 1; nuclear alterations and hyperplasia scored as 2; necrosis scored as 3; and malformation scored as 4.

2.5. Quantitative Real-Time RT-PCR Assay

After exposure, larvae (fifty individuals per sample, n = 3/group) were randomly selected to quantitatively analyze the transcriptional levels of genes, including *sod*, *cat*, *gpx1a*, *gpx4a*, *gstt1a*, *fosab*, *fosb*, *egr2a*, *egr2b*, *egr4*, *tnf-α*, *il-6*, *tp53*, *casp3*, *rrm2*, and *β-Actin*. Total RNA of each sample was extracted, reverse transcribed, and determined by real-time polymerase chain reaction. Detailed information is described in the Supporting Information, and all primer sequences are shown in Table S2.

2.6. Statistical Analysis

Data were analyzed using SPSS 26 software (SPSS, Inc., Chicago, IL, USA). The normality of the data was determined using Shapiro-Wilk's test. Homogeneity of variance was assessed using Levine's test. The survival curve was established by GraphPad Prism 8.0 (GraphPad Software, San Diego, CA, USA), and the differences were estimated using the log rank test. The differences in the mortality observed at different time points between the control group and the exposure groups were determined using two-way ANOVA. One-way ANOVA followed by Tukey's test was used to determine the differences in the terminal mortalities, values of behavioral parameters, and transcriptional levels of genes among different groups. Student's t test was used to analyze the differences in the data between two groups. The correlation analysis among different parameters of fish in different groups was conducted by calculating Pearson's correlation coefficients. All data are expressed as the mean ± standard deviation (SD). Statistical significance was considered as p values less than 0.05 (95% confidence interval).

3. Results

3.1. Characterization Map, Zeta Potential, and Analysis of MPs

Monomers of PS and PVC MPs appeared as smooth and intact spheres (Figure S1). The average sizes of PS and PVC used in this work were 6.998 and 3.760 μm, respectively. As listed in Table S3, the zeta potentials of PS and PVC MPs were ± 45 and ± 20 (pH = 7 and 25 °C), respectively.

3.2. Effects of Combined MPs and METH Exposure on the Mortality of Zebrafish Larvae

During the 10 d exposure, the death of larvae in the control group occurred until 5 d, while exposure to individual PS or PVC induced constant wastage of the community

($p < 0.01$ for PS and PVC, Figure 1b). Compared with the individual exposure groups, there were no significant differences observed in the joint exposure groups (Figure 1c,d). Considering the effect of exposure duration, the daily changes in mortality from different groups were analyzed (see Figure 1e). The slope of the changing curve of mortality of the PS groups was greater than that of the PVC group. According to the terminal lethal results (Figure 1f), the total mortality of MPs combined with METH was lower than that of single MPs exposure and markedly greater than that of the control. Meanwhile, there was a significant difference in total mortality between the PVC250 and PS250 groups ($p < 0.01$).

3.3. Effects of Combined MPs and METH Exposure on the Locomotor Behavior of Zebrafish Larvae

Unexposed zebrafish exhibited a characteristic behavior of swimming along the walls of the container (the control in Figure 2a). Notably, treatment with MPs or METH markedly induced an abnormal behavioral phenotype in fishes, characterized as locomotor activity syndrome with lateralized circling behaviors and postural asymmetry. As shown in Figure 2a, the animals from the PS group and PVC group showed decreased swimming activity compared to the control, while the addition of METH alleviated this effect. The immobility duration of larvae exposed to individual PS and the mixture of PS and METH increased significantly compared with the control ($p < 0.05$), featuring a reverse "U" shape (Figure 2b). The peak was observed in the PS250 group. For the mean velocity (Figure 2c), individual exposure to PS reduced motility, but joint exposure to PS and METH at 250 μg L^{-1} strikingly stimulated movement and then returned to normal levels (the control) with METH up to 500 μg L^{-1}. The changes in the turn angle of fishes from the PS group showed a concentration-dependent manner (Figure 2d). Notably, the immobility duration of zebrafish from the PVC group showed the opposite trajectory as a "U" shape, in which the culmination was observed from the individual PVC group (Figure 2b). Treatment with PVC markedly suppressed the locomotion of fishes ($p < 0.05$), while the values increased with METH addition (Figure 2c). Meanwhile, the alterations in the turn angle from the PVC groups showed a similar pattern to the PS group (Figure 2d).

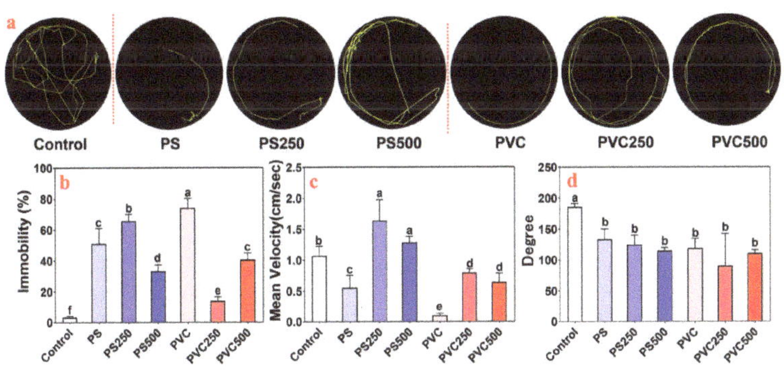

Figure 2. Behavioral changes of zebrafish larvae in different groups after 10-day exposure. (**a**) Representative swimming trajectories of zebrafish larvae. (**b–d**) The mobility rate, mean velocity, and degree of zebrafish larvae of the group exposed to PS MPs, PS MPs + METH, PVC MPs and PVC MPs + METH. Distinct letters denote statistically significant differences among the respective groupings.

3.4. Histopathological Effects of Combined MPs and METH Exposure on Zebrafish Larvae

Obvious particulate-like sediments were found in the enteric cavity of fish from the exposure groups (red arrow) (Figure 3b–g) compared with the control (Figure 3a), suggesting that MPs are foraged by larvae through the digestive tract. Meanwhile, a variety of histopathological changes were observed in the fish from different exposure groups, including cyrtosis (black arrow), pleural effusion (blue arrow), and inflammatory cell infiltration (green arrow) (Figure 3b–g) in comparison with the control (Figure 3a).

According to the histopathological score evaluation (Figure 3h), combined exposure to PVC MPs and METH at 500 μg L^{-1} induced the most significant changes, followed by PS MPs + METH at 250 μg L^{-1}, and PS MPs showed slightly greater effects than PVC MPs.

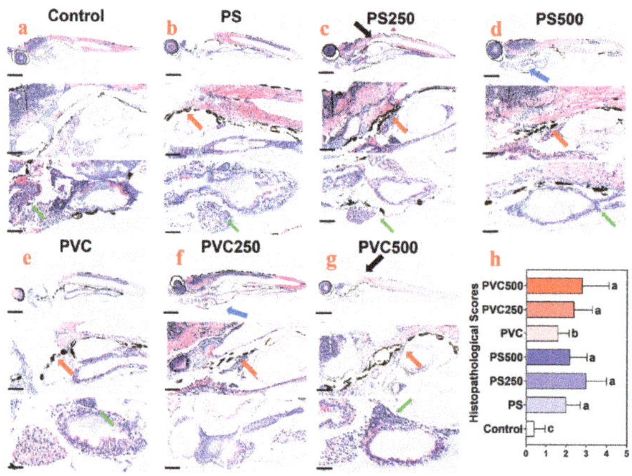

Figure 3. Histopathological changes of zebrafish larvae observed in the control group (**a**), PS MPs group (**b**), PS MPs + METH (**c**,**d**), PVC MPs (**e**), and PVC MPs + METH (**f**,**g**), as well as the histopathological scores in different groups (**h**). Pleural effusion (blue arrow), cyrtosis (black arrow), inflammatory cell infiltration (green arrow) and intestinal contents (red arrows). Scale bar = 100 μm. The rank evaluations on the histopathological changes of zebrafish (**h**). Distinct letters denote statistically significant differences among the respective groupings.

3.5. Effect of Combined MPs and METH Exposure on Gene Expression in Zebrafish Larvae

The relative expression levels of the *sod*, *cat*, *gpx1a*, *gpx4a*, and *gstt1a* genes involved in oxidative stress were determined in this work (Figure 4a). Individual exposure to MPs generally significantly induced the expression of antioxidant enzyme-related genes. The fold changes were 2.14-fold to 34.95-fold, respectively. After the addition of METH, their relative expression levels showed a significant decrease (−0.89-fold to 14.75-fold).

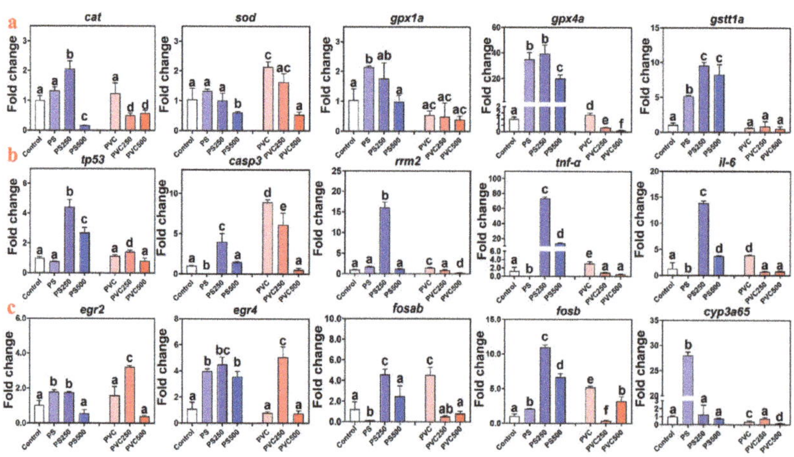

Figure 4. Relative gene expression levels of zebrafish larvae in different groups involved into the pathways including oxidative stress pathways (**a**), apoptosis (**b**), and xenobiotic metabolism (**c**). All the data were expressed as mean ± S.D. Distinct letters denote statistically significant differences among the respective groupings.

The relative expression levels of the *tnf-α*, *il-6*, *casp3*, *rrm2*, and *tp53* genes involved in the p53 signaling pathway responsible for apoptosis were determined in this work (Figure 4b). The exposure of zebrafish larvae to MPS and the blend of MPs and METH resulted in an inverted "U" shaped change in the expression levels of immune-related genes, with the maximum value observed in the PS250 group. The fold changes were 4.40-fold to 73.53-fold, respectively. In the PVC exposure group, the crest was observed in the individual PVC exposure group (1.13-fold for tp53, 8.92-fold for *casp3*, 1.46-fold for *rrm2*, 3.09-fold for *tnf-α*, and 3.87-fold for *il-6*).

Given the pronounced histopathological changes found in fish from the exposure groups, the mRNA transcriptional levels of larvae development- and metabolism-associated genes (i.e., *egr2*, *egr4*, *fosab*, *fosb*, and *cyp3a65*) were analyzed. As shown in Figure 4c, all the changes in PS MPs groups showed the reverse "U" shape, with the maximum values (except *cyp3a65*) found in the PS250 groups, and the fold-change ranged from 1.76-fold to 107.84-fold. For *cyp3a65*, the peak was observed in the PS group (30-fold). In the PVC groups, the transcriptional levels of the genes *egr2*, *egr4*, *fosab*, and *fosb* were upregulated 3.22-fold to 5.20-fold, but *cyp3a65* was downregulated by 0.16-fold.

3.6. Correlation Analysis between the Different Biomarkers of Fish in Different Groups

To further elucidate the interaction relationships among lethal effects, behavioral disorders, histopathological damage, and genetic expression changes, Pearson's correlation coefficients were calculated with the genes on the *y*-axis and physiological indicators on the *x*-axis (Figure 5a). Mortality was positively correlated with most of the genes, and immobility was positively related to the genes *cat*, *rrm2*, *il-6*, and *fosb* ($p < 0.05$). Mean velocity showed covariation relationships with the genes involved in oxidative stress and the P53 signaling pathway. Furthermore, a chord diagram was established to illuminate the cross-talk among different parameters (Figure 5b). ROS were strongly linked to apoptosis and xenobiotic metabolism and moderately associated with behavioral functions and mortality, indicating that they might be the primary events for the adverse outcomes in zebrafish posed by MPs and METH. Otherwise, apoptosis and xenobiotic metabolism showed a correlative relationship. Histopathological changes were weakly associated with apoptosis, behavioral functions, and xenobiotic metabolism.

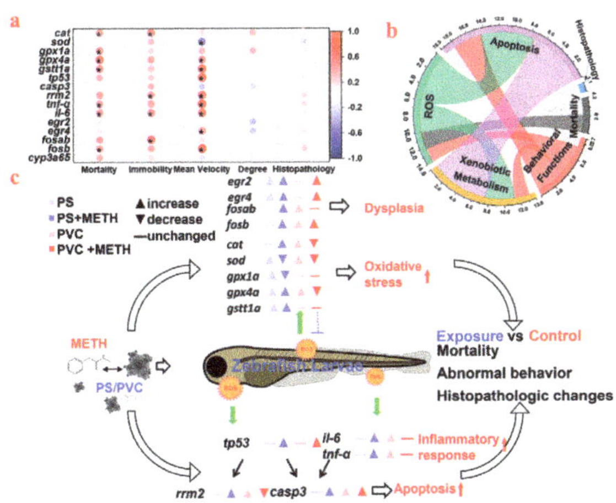

Figure 5. The interaction relationships among lethal effects, behavioral disorders, histopathological damage, and genetic expression changes (Pearson's correlation coefficients) (**a**); a chord diagram is employed to visualize and elucidate the cross−talk among various parameters (**b**); a pattern diagram depicting alterations in gene expression profiles in zebrafish larvae following the ingestion of polypropylene microplastics (PP MPs) and methamphetamine in vivo (**c**). * $p < 0.05$.

4. Discussion

4.1. The Effects on the Ecological Function Associated Indicators of Zebrafish Larvae by MPs

Much existing evidence, such as aberrant swimming behavior, glycolipid metabolism disorder, and immune responses, has revealed the toxicity of PS to fish [9–12]. The MPs utilized in this study have a diameter of 10 μm. Previous research by Lin et al. indicated that they are unable to enter the chorionic pore of zebrafish embryos [46]. However, it is widely acknowledged that MPs adsorbed on the surface of the chorionic pore can still exert an impact on zebrafish embryos (e.g., increase in incubation time and increased heart rate) [47,48]. After hatching of zebrafish larvae, MPs enter the larvae by being mistaken for food or carried into the larvae by water, thus causing toxicity in behavior and development [10]. Similarly, we proved the toxicity of 20 mg L^{-1} PS on the behavior and development of zebrafish larvae, indicating that ecological risks should be considered (Figure 5c). Parallel to previous assessments on the toxicity for zebrafish embryos induced by PVC MPs [29], intestinal damage and oxidative stress in the animals were observed. Otherwise, the increase in mortality and associated deformations were first reported in this work. Considering the cocktails of MPs and emerging organic pollutants in aquatic systems, assessment of synergetic toxicity is imperative. Since the adsorption of METH by MPs was evidenced [40], the different toxicological effects posed by individual PS and PVC MPs and joints with METH were investigated.

4.2. Oxidative Stress

Fish, especially during their early life stages, are vulnerable to oxidative stress [49]. Recent research has emphasized that exposure to MPs is a major contributor to oxidative stress in zebrafish [50]. This study specifically investigated the impact of MPs on oxidative stress in zebrafish larvae. The results demonstrated that a single exposure to PS MPs significantly influenced oxidative stress by upregulating the expression of the *gpx1a*, *gpx4a*, and cat genes. These genes play a critical role in converting hydrogen peroxide into water, thereby maintaining cellular oxidative balance [29]. Notably, single exposure to PS MPs did not significantly induce *sod* expression, whereas PVC exposure led to a significant increase in *sod* expression but did not affect cat and *gpx1a* expression. This suggests that the impact of MPs on oxidative stress is contingent upon the type of MPs present, with PS MPs inducing higher levels of oxidative stress than PVC MPs. This study revealed that the addition of METH resulted in a significant reduction in the expression of antioxidant enzyme genes in zebrafish larvae, particularly at a concentration of 500 μg L^{-1} METH. The results of the correlation analysis confirmed a strong positive correlation between the expression of antioxidant enzymes and the mortality rate (Figure 5a). This implies that oxidative stress is the main cause of mortality in zebrafish larvae, and METH and MPs exhibit an antagonistic relationship in terms of oxidative stress and lethal toxicity. These findings shed light on the intricate interactions between MPs and substances like METH in influencing oxidative stress responses and mortality in zebrafish larvae.

Oxidative stress is closely related to inflammatory reactions [51]. Due to the ROS induced by MPs exposure, the p53 pathway has been shown to be significantly upregulated, leading to cell apoptosis and DNA damage [52]. The histopathological examination of zebrafish larvae showed a significant inflammatory reaction, including inflammatory cell infiltration and pleural edema. Meanwhile, suspected levels of MPs were also found in the digestive tract, which have been confirmed to cause mucosal damage and inflammation [53]. Genes *il-6* and *tnf- α* are inflammatory factors, and their relative expression can represent the degree to which the body is affected by inflammation Interestingly, we found that exposure to PVC significantly up-regulated the expression of associated genes, while there were no marked changes observed after exposure to PS alone. Addition with METH mited the changes for PVC but catalyzed the effects for PS. At the same time, the inflammatory response also activated the p53 pathway related to cell apoptosis. By examining downstream genes [54], including *tp53*, *casp3*, and *rrm2*, a positive correlation was revealed between their expression levels and the expression level of inflammation-related genes. The significant increase in *casp3* expression indicates the induction of apoptosis, leading to the death of zebrafish larvae [55]. In addition, changes in *rrm2* expression indicate DNA damage, as organisms attempt to regulate and repair damaged DNA [56]. The potential reasons may be that the adsorption capacity of PS and polar organic pollutants is significantly higher than that of PVC [30]. Meanwhile, the zeta potential of PVC determines that its alternating stability in water is not as good as that of PS, and it is more prone to agglomeration, thereby reducing the adsorption of METH [57]. As a result, PS might carry much more METH into the zebrafish body than PVC, and it leads to more severe oxidative stress phenomena. More experiments should be conducted in the future to test this hypothesis. These findings highlight the potential synergistic effects of MPs and emerging organic pollutants in aquatic environments and the need for continued research to better understand their mechanisms of toxicity and ecological impacts.

4.3. Ecological Implications

Behavioral functions are important indicators for assessing the neurotoxicity of pollutants on aquatic organisms, including fishes. Changes in normal behavior can have significant consequences at the ecosystem level [58]. Swimming ability, for instance, directly interferes with the success of fish in avoiding predators and actively feeding [59]. In this study, zebrafish larvae exposed to MPs alone showed an aberrant swimming trajectory compared with the control, which was consistent with the results found in a previous publication [10]. When exposed to a combination of MPs and METH, a hormetic response was observed, with stronger effects at a concentration of 250 $\mu g\ L^{-1}$ and alleviated effects at 500 $\mu g\ L^{-1}$. The observed phenomenon may be attributed to the primary stimulatory effects of METH at low concentrations and counteractive effects at higher doses. In line with findings in existing literature, *Oryzias latipes* display a significant increase in movement at lower concentrations of METH exposure. However, beyond a certain threshold concentration, this heightened movement begins to decrease, aligning with the results of our study. The potential reason for this change in our study could be the variation in METH adsorption capacity among different types of microplastics, resulting in differences in the concentrations to which zebrafish larvae are exposed [43]. Furthermore, the turn angle, which represents the behavior of exploring the environment, decreased in a concentration-dependent manner with exposure to MPs and METH, indicating an increased risk of predation for the exposed fishes [58]. Exiting studies have revealed that zebrafish exhibit reduced activity accompanied by inflammatory responses [60]. Similarly, the results of correlation analysis showed a strong correlation between inflammation-related genes and immobility, which may be due to the "anxiety" phenomenon induced by inflammation in zebrafish larvae [61]. In addition, oxidative stress-related genes also exhibited covariation relationships with behavioral changes in zebrafish larvae. Chen et al. suggested that oxidative stress could affect brain neurodevelopment, thus influencing the behavioral changes in zebrafish larvae [62], which might be one of the reasons for the abnormal swimming speed observed in zebrafish larvae.

In addition, MPs may interfere with the early development of zebrafish larvae, thereby impairing their motor ability [10], especially the neurodevelopmental toxicity [63] caused by them. In this study, the relative expression of genes related to the early development of zebrafish larvae was also studied; among these genes, *fosab* and *fosb* are translation factors in zebrafish brain nerves that act as natural markers and participate in early embryonic development [64]. In this study, the amount of gene expression was also the same as the type of MPs. In addition, there were diametrically opposite changes after adding METH. Such abnormal expression may lead to abnormal brain development and consequently affect the response of zebrafish larvae to external stimuli. In addition, early growth response (egr) factors have been elucidated to be related to the proliferation and differentiation of zebrafish in early life [65]. Recent studies by Lee et al. have shown that exposure to PS MPs significantly downregulates the expression of mouse EGR family genes [66]. Consistent with these findings, in this study, similar changes in *egr* factor expression were observed in zebrafish exposed to PS and PVC MPs. However, the addition of METH alleviated the adverse effects on *egr* factors, indicating a potential beneficial effect of METH combination therapy in alleviating the developmental delay caused by MPs. In particular, *egr2* has been found to regulate the expression of *cyp26b1* through retinoic acid signaling, and *cyp26b1*, which is mainly expressed in osteoblasts, has been proven to cause head and face defects and axial bone abnormalities in zebrafish [67,68]. Therefore, the interference of PS and PVC MPs on the functions of the *egr* factor and *c-fos* factor in zebrafish larvae may be the cause of movement and morphological changes in the exposed group. Further research is needed to fully understand the potential mechanism of interaction between the expression of MPs, METH, and key regulators in zebrafish larvae.

4.4. Perspectives for Joint Toxicity of METH and PS/PVC MPs for Aquatic Organisms

Overall, the study indicates that PS and PVC MPs have distinct toxic effects on zebrafish larvae when exposed individually. PS MPs exposure significantly induces the expression of *cat*, *gpx1a*, and *gpx4a*, which are related to oxidative stress, while PVC MPs exposure mainly leads to a significant induction of *sod*. This suggests that PS exposure results in a more pronounced increase in oxidative stress-related enzymes compared to PVC exposure, and the expression of these oxidative stress-related enzymes is significantly correlated with the mortality rate. Moreover, when combined with METH, joint exposure to PS MPs and METH activates the p53 signaling pathway and enhances the expression of inflammation-related factors. In contrast, combined exposure to PVC MPs and METH inhibited their expression. These findings may be attributed to the ability of MPs to carry METH and the inherent differences in the toxicity of MPs themselves. It is worth noting that even though both types of MPs are categorized as <10 microns in size, there may be variations in particle size between PS and PVC MPs, potentially contributing to differences in their toxicity based on previous research [69]. In conclusion, further research involving joint toxicity experiments on MPs and psychotropic drugs is warranted to gain a better understanding of the intricate relationship between MPs and psychotropic drugs. Such investigations will help uncover the mechanisms underlying their combined effects and shed light on the potential ecological implications of these interactions.

5. Conclusions

In summary, the characteristics of the synergetic toxicity of MPs and METH was highly dependent on the type of MPs. Individual exposure to PS was more toxic than PVC based on the changes of the mortality and the behavioral functions. Meanwhile, joint exposure by PS and PVC with METH featured varying patterns, which might attribute to the different adsorption capacity of this two MPs. The cocktail of PS and METH was more likely to pose adverse effects on zebrafish larvae. This study, for the first time, estimated the different toxicity between PS and PVC MPs and the joint effects posed by MPs and METH. The results demonstrated the necessary for long-term monitoring of the pollutions in aquatic environments.

Supplementary Materials: The following supporting information can be downloaded at: https://www.mdpi.com/article/10.3390/toxics12010009/s1, Figure S1: Electron microscope scans of PS and PVC microplastics, Monomer of PS (a–c) and PVC (d–f) MPs showed as smooth and intact sphere; Table S1: The measured and nominal concentrations of METH in exposure solution; Table S2: Primer sequences for the genes tested in the present study; Table S3: ZETA potential of PS and PVC microplastics (pH = 7, 25 °C); Table S4: Relative expression change fold of genes in PS and METH exposure combination. Table S5: Relative expression change fold of genes in PVC and METH exposure combination.

Author Contributions: J.X.: Investigation, Writing—Original Draft, Writing—Review & Editing W.Y. and D.W.: Formal analysis, Data Curation. J.L., Z.W., and C.L.: Conceptualization, Supervision, Resources, Writing—Review & Editing. All authors have read and agreed to the published version of the manuscript.

Funding: This study was financially supported by the Natural Science Foundation of Jiangsu Province (No. BK20210376) and the Ningbo Public Welfare Technology Plan Project (No. 2022S118).

Institutional Review Board Statement: All animal studies were performed in accordance with the Guidelines for Animal Experiments of Hohai University, which meet the ethical guidelines for experimental animals in China.

Data Availability Statement: Data are contained within the article.

Acknowledgments: The author sincerely thanks his graduate supervisor and beloved bicycle.

Conflicts of Interest: The authors declare no conflict of interest.

References

1. Tang, Y.; Liu, Y.; Chen, Y.; Zhang, W.; Zhao, J.; He, S.; Yang, C.; Zhang, T.; Tang, C.; Zhang, C.; et al. A review: Research progress on microplastic pollutants in aquatic environments. *Sci. Total Environ.* **2021**, *766*, 142572. [CrossRef] [PubMed]
2. Europe, P. An analysis of European Plastics Production, Demand and Waste Data, Plastics-The Facts 2019. Available online: https://plasticseurope.org/knowledge-hub/plastics-the-facts-2019/ (accessed on 10 March 2023).
3. Arthur, C.; Baker, J.; Bamford, H. In Proceedings of the International Research Workshop on the Occurrence, Effects, and Fate of Microplastic Marine Debris, University of Washington Tacoma, Tacoma, WA, USA, 9–11 September 2008.
4. Su, L.; Xue, Y.; Li, L.; Yang, D.; Kolandhasamy, P.; Li, D.; Shi, H. Microplastics in Taihu Lake, China. *Environ. Pollut.* **2016**, *216*, 711–719. [CrossRef] [PubMed]
5. Zhao, S.; Zhu, L.; Wang, T.; Li, D. Suspended microplastics in the surface water of the Yangtze Estuary System, China: First observations on occurrence, distribution. *Mar. Pollut. Bull.* **2014**, *86*, 562–568. [CrossRef] [PubMed]
6. Mani, T.; Hauk, A.; Walter, U.; Burkhardt-Holm, P. Microplastics profile along the Rhine River. *Sci. Rep.* **2015**, *5*, 17988. [CrossRef] [PubMed]
7. Wang, Z.; Lin, T.; Chen, W. Occurrence and removal of microplastics in an advanced drinking water treatment plant (ADWTP). *Sci. Total Environ.* **2020**, *700*, 134520. [CrossRef] [PubMed]
8. Lusher, A.; Welden, N.; Sobral, P.; Cole, M. Sampling, isolating and identifying microplastics ingested by fish and invertebrates. In *Analysis of Nanoplastics and Microplastics in Food*; CRC Press: Boca Raton, FL, USA, 2020; pp. 119–148.
9. Zhang, C.; Chen, X.; Wang, J.; Tan, L. Toxic effects of microplastic on marine microalgae Skeletonema costatum: Interactions between microplastic and algae. *Environ. Pollut.* **2017**, *220*, 1282–1288. [CrossRef] [PubMed]
10. Qiang, L.; Cheng, J. Exposure to microplastics decreases swimming competence in larval zebrafish (*Danio rerio*). *Ecotoxicol. Environ. Saf.* **2019**, *176*, 226–233. [CrossRef] [PubMed]
11. Hamed, M.; Soliman, H.A.M.; Osman, A.G.M.; Sayed, A.E.-D.H. Assessment the effect of exposure to microplastics in Nile Tilapia (*Oreochromis niloticus*) early juvenile: I. blood biomarkers. *Chemosphere* **2019**, *228*, 345–350. [CrossRef]
12. De Marco, G.; Conti, G.O.; Giannetto, A.; Cappello, T.; Galati, M.; Iaria, C.; Pulvirenti, E.; Capparucci, F.; Mauceri, A.; Ferrante, M.; et al. Embryotoxicity of polystyrene microplastics in zebrafish Danio rerio. *Environ. Res.* **2022**, *208*, 112552. [CrossRef]
13. Corami, F.; Rosso, B.; Morabito, E.; Rensi, V.; Gambaro, A.; Barbante, C. Small microplastics (<100 μm), plasticizers and additives in seawater and sediments: Oleo-extraction, purification, quantification, and polymer characterization using Micro-FTIR. *Sci. Total Environ.* **2021**, *797*, 148937. [CrossRef]
14. Koelmans, A.A.; Redondo-Hasselerharm, P.E.; Nor, N.H.M.; de Ruijter, V.N.; Mintenig, S.M.; Kooi, M. Risk assessment of microplastic particles. *Nat. Rev. Mater.* **2022**, *7*, 138–152. [CrossRef]
15. Liu, C.; Huang, W.; Gu, Y.; Rao, Z. Recent advances in high value added reuse of waste polystyrene in environment and energy. *CIESC J.* **2020**, *71*, 2956–2972. [CrossRef]
16. Zhang, Z.; Gao, S.-H.; Luo, G.; Kang, Y.; Zhang, L.; Pan, Y.; Zhou, X.; Fan, L.; Liang, B.; Wang, A. The contamination of microplastics in China's aquatic environment: Occurrence, detection and implications for ecological risk. *Environ. Pollut.* **2022**, *296*, 118737. [CrossRef] [PubMed]
17. Strom, S.L.; Fredrickson, K.A.; Bright, K.J. Microzooplankton in the coastal Gulf of Alaska: Regional, seasonal and interannual variations. *Deep Sea Res. Part II Top. Stud. Oceanogr.* **2019**, *165*, 192–202. [CrossRef]
18. Zhang, F.; Wang, Z.; Wang, S.; Fang, H.; Wang, D. Aquatic behavior and toxicity of polystyrene nanoplastic particles with different functional groups: Complex roles of pH, dissolved organic carbon and divalent cations. *Chemosphere* **2019**, *228*, 195–203. [CrossRef] [PubMed]
19. Liu, Y.; Wang, Z.; Wang, S.; Fang, H.; Ye, N.; Wang, D. Ecotoxicological effects on Scenedesmus obliquus and Danio rerio Co-exposed to polystyrene nano-plastic particles and natural acidic organic polymer. *Environ. Toxicol. Pharmacol.* **2019**, *67*, 21–28. [CrossRef] [PubMed]
20. Thanh Truc, N.T.; Lee, C.-H.; Lee, B.-K.; Mallampati, S.R. Development of hydrophobicity and selective separation of hazardous chlorinated plastics by mild heat treatment after PAC coating and froth flotation. *J. Hazard. Mater.* **2017**, *321*, 193–202. [CrossRef]
21. Xia, X.; Sun, M.; Zhou, M.; Chang, Z.; Li, L. Polyvinyl chloride microplastics induce growth inhibition and oxidative stress in Cyprinus carpio var. larvae. *Sci. Total Environ.* **2020**, *716*, 136479. [CrossRef]

22. Wang, Z.; Mao, K.; Du, W.; Cai, M.; Zhang, Z.; Li, X. Diluted concentrations of methamphetamine in surface water induce behavior disorder, transgenerational toxicity, and ecosystem-level consequences of fish. *Water Res.* **2020**, *184*, 116164. [CrossRef]
23. Koelmans, A.A.; Bakir, A.; Burton, G.A.; Janssen, C.R. Microplastic as a Vector for Chemicals in the Aquatic Environment: Critical Review and Model-Supported Reinterpretation of Empirical Studies. *Environ. Sci. Technol.* **2016**, *50*, 3315–3326. [CrossRef]
24. Rochman, C.M.; Kurobe, T.; Flores, I.; Teh, S.J. Early warning signs of endocrine disruption in adult fish from the ingestion of polyethylene with and without sorbed chemical pollutants from the marine environment. *Sci. Total Environ.* **2014**, *493*, 656–661 [CrossRef] [PubMed]
25. Granby, K.; Rainieri, S.; Rasmussen, R.R.; Kotterman, M.J.J.; Sloth, J.J.; Cederberg, T.L.; Barranco, A.; Marques, A.; Larsen, B.K The influence of microplastics and halogenated contaminants in feed on toxicokinetics and gene expression in European seabass (*Dicentrarchus labrax*). *Environ. Res.* **2018**, *164*, 430–443. [CrossRef] [PubMed]
26. Qu, H.; Ma, R.; Wang, B.; Yang, J.; Duan, L.; Yu, G. Enantiospecific toxicity, distribution and bioaccumulation of chiral antidepressant venlafaxine and its metabolite in loach (*Misgurnus anguillicaudatus*) co-exposed to microplastic and the drugs. *J. Hazard. Mater.* **2019**, *370*, 203–211. [CrossRef] [PubMed]
27. Wang, Y.; Zhang, M.; Ding, G.; Shi, H.; Cong, Y.; Li, Z.; Wang, J. Polystyrene microplastics alleviate adverse effects of benzo[a]pyrene on tissues and cells of the marine mussel, Mytilus galloprovincialis. *Aquat. Toxicol.* **2023**, *256*, 106430. [CrossRef] [PubMed]
28. Syberg, K.; Nielsen, A.; Khan, F.R.; Banta, G.T.; Palmqvist, A.; Jepsen, P.M. Microplastic potentiates triclosan toxicity to the marine copepod Acartia tonsa (Dana). *J. Toxicol. Environ. Health Part A* **2017**, *80*, 1369–1371. [CrossRef] [PubMed]
29. Wang, H.; Wang, Y.; Wang, Q.; Lv, M.; Zhao, X.; Ji, Y.; Han, X.; Wang, X.; Chen, L. The combined toxic effects of polyvinyl chloride microplastics and di(2-ethylhexyl) phthalate on the juvenile zebrafish (*Danio rerio*). *J. Hazard. Mater.* **2022**, *440*, 129711. [CrossRef]
30. Li, J.; Zhang, K.; Zhang, H. Adsorption of antibiotics on microplastics. *Environ. Pollut.* **2018**, *237*, 460–467. [CrossRef] [PubMed]
31. Fu, L.; Li, J.; Wang, G.; Luan, Y.; Dai, W. Adsorption behavior of organic pollutants on microplastics. *Ecotoxicol. Environ. Saf.* **2021**, *217*, 112207. [CrossRef]
32. Zhu, Z.-L.; Wang, S.-C.; Zhao, F.-F.; Wang, S.-G.; Liu, F.-F.; Liu, G.-Z. Joint toxicity of microplastics with triclosan to marine microalgae Skeletonema costatum. *Environ. Pollut.* **2019**, *246*, 509–517. [CrossRef]
33. Rubin, A.E.; Zucker, I. Interactions of microplastics and organic compounds in aquatic environments: A case study of augmented joint toxicity. *Chemosphere* **2022**, *289*, 133212. [CrossRef]
34. Chen, Y.; Wisner, A.S.; Schiefer, I.T.; Williams, F.E.; Hall, F.S. Methamphetamine-induced lethal toxicity in zebrafish larvae. *Psychopharmacology* **2022**, *239*, 3833–3846. [CrossRef] [PubMed]
35. Lin, A.Y.-C.; Wang, X.-H.; Lin, C.-F. Impact of wastewaters and hospital effluents on the occurrence of controlled substances in surface waters. *Chemosphere* **2010**, *81*, 562–570. [CrossRef]
36. Paciuszkiewicz, K.; Ryan, M.; Wright, I.A.; Reynolds, J.K. Variations in Illicit Compound Discharged from Treated Wastewater. *Water* **2019**, *11*, 1071. [CrossRef]
37. Mastroianni, N.; Bleda, M.J.; López de Alda, M.; Barceló, D. Occurrence of drugs of abuse in surface water from four Spanish river basins: Spatial and temporal variations and environmental risk assessment. *J. Hazard. Mater.* **2016**, *316*, 134–142. [CrossRef] [PubMed]
38. Liao, P.-H.; Hwang, C.-C.; Chen, T.-H.; Chen, P.-J. Developmental exposures to waterborne abused drugs alter physiological function and larval locomotion in early life stages of medaka fish. *Aquat. Toxicol.* **2015**, *165*, 84–92. [CrossRef] [PubMed]
39. Horký, P.; Grabic, R.; Grabicová, K.; Brooks, B.W.; Douda, K.; Slavík, O.; Hubená, P.; Sancho Santos, E.M.; Randák, T. Methamphetamine pollution elicits addiction in wild fish. *J. Exp. Biol.* **2021**, *224*, jeb242145. [CrossRef] [PubMed]
40. Qu, H.; Ma, R.; Barrett, H.; Wang, B.; Han, J.; Wang, F.; Chen, P.; Wang, W.; Peng, G.; Yu, G. How microplastics affect chiral illicit drug methamphetamine in aquatic food chain? From green alga (*Chlorella pyrenoidosa*) to freshwater snail (*Cipangopaludian cathayensis*). *Environ. Int.* **2020**, *136*, 105480. [CrossRef]
41. Iwamatsu, T. Stages of normal development in the medaka Oryzias latipes. *Mech. Dev.* **2004**, *121*, 605–618. [CrossRef]
42. Xue, Y.-H.; Feng, L.-S.; Xu, Z.-Y.; Zhao, F.-Y.; Wen, X.-L.; Jin, T.; Sun, Z.-X. The time-dependent variations of zebrafish intestine and gill after polyethylene microplastics exposure. *Ecotoxicology* **2021**, *30*, 1997–2010. [CrossRef]
43. Wang, Z.; Dai, S.; Wang, J.; Du, W.; Zhu, L. Assessment on chronic and transgenerational toxicity of methamphetamine to Caenorhabditis elegans and associated aquatic risk through toxicity indicator sensitivity distribution (TISD) analysis. *Environ. Pollut.* **2021**, *288*, 117696. [CrossRef]
44. Gulyás, M.; Bencsik, N.; Pusztai, S.; Liliom, H.; Schlett, K. AnimalTracker: An ImageJ-Based Tracking API to Create a Customized Behaviour Analyser Program. *Neuroinformatics* **2016**, *14*, 479–481. [CrossRef] [PubMed]
45. Bernet, D.; Schmidt, H.; Meier, W.; Burkhardt-Holm, P.; Wahli, T. Histopathology in fish: Proposal for a protocol to assess aquatic pollution. *J. Fish Dis.* **1999**, *22*, 25–34. [CrossRef]
46. Lin, S.; Zhao, Y.; Nel, A.E.; Lin, S. Zebrafish: An in vivo model for nano EHS studies. *Small* **2013**, *9*, 1608–1618. [CrossRef] [PubMed]
47. Duan, Z.; Duan, X.; Zhao, S.; Wang, X.; Wang, J.; Liu, Y.; Peng, Y.; Gong, Z.; Wang, L. Barrier function of zebrafish embryonic chorions against microplastics and nanoplastics and its impact on embryo development. *J. Hazard. Mater.* **2020**, *395*, 122621. [CrossRef] [PubMed]

48. Mu, X.; Qi, S.; Liu, J.; Yuan, L.; Huang, Y.; Xue, J.; Qian, L.; Wang, C.; Li, Y. Toxicity and behavioral response of zebrafish exposed to combined microplastic and bisphenol analogues. *Environ. Chem. Lett.* **2022**, *20*, 41–48. [CrossRef]
49. Félix, L.M.; Vidal, A.M.; Serafim, C.; Valentim, A.M.; Antunes, L.M.; Campos, S.; Matos, M.; Monteiro, S.M.; Coimbra, A.M. Ketamine-induced oxidative stress at different developmental stages of zebrafish (*Danio rerio*) embryos. *RSC Adv.* **2016**, *6*, 61254–61266. [CrossRef]
50. Lei, L.; Wu, S.; Lu, S.; Liu, M.; Song, Y.; Fu, Z.; Shi, H.; Raley-Susman, K.M.; He, D. Microplastic particles cause intestinal damage and other adverse effects in zebrafish Danio rerio and nematode Caenorhabditis elegans. *Sci. Total Environ.* **2018**, *619–620*, 1–8. [CrossRef] [PubMed]
51. Lu, K.; Qiao, R.; An, H.; Zhang, Y. Influence of microplastics on the accumulation and chronic toxic effects of cadmium in zebrafish (*Danio rerio*). *Chemosphere* **2018**, *202*, 514–520. [CrossRef]
52. Umamaheswari, S.; Priyadarshinee, S.; Kadirvelu, K.; Ramesh, M. Polystyrene microplastics induce apoptosis via ROS-mediated p53 signaling pathway in zebrafish. *Chem. Biol. Interact.* **2021**, *345*, 109550. [CrossRef]
53. Qiao, R.; Deng, Y.; Zhang, S.; Wolosker, M.B.; Zhu, Q.; Ren, H.; Zhang, Y. Accumulation of different shapes of microplastics initiates intestinal injury and gut microbiota dysbiosis in the gut of zebrafish. *Chemosphere* **2019**, *236*, 124334. [CrossRef]
54. Perfettini, J.-L.; Roumier, T.; Castedo, M.; Larochette, N.; Boya, P.; Raynal, B.; Lazar, V.; Ciccosanti, F.; Nardacci, R.; Penninger, J.; et al. NF-κB and p53 Are the Dominant Apoptosis-inducing Transcription Factors Elicited by the HIV-1 Envelope. *J. Exp. Med.* **2004**, *199*, 629–640. [CrossRef] [PubMed]
55. Yabu, T.; Kishi, S.; Okazaki, T.; Yamashita, M. Characterization of zebrafish caspase-3 and induction of apoptosis through ceramide generation in fish fathead minnow tailbud cells and zebrafish embryo. *Biochem. J.* **2001**, *360*, 39–47. [CrossRef] [PubMed]
56. Niida, H.; Katsuno, Y.; Sengoku, M.; Shimada, M.; Yukawa, M.; Ikura, M.; Ikura, T.; Kohno, K.; Shima, H.; Suzuki, H. Essential role of Tip60-dependent recruitment of ribonucleotide reductase at DNA damage sites in DNA repair during G1 phase. *Genes Dev.* **2010**, *24*, 333–338. [CrossRef] [PubMed]
57. Xiao, Y.; Li, D.; Fan, H.; Li, X.; Gu, Z.; Zhang, X. Preparation of nano-HA/PLA composite by modified-PLA for controlling the growth of HA crystals. *Mater. Lett.* **2007**, *61*, 59–62. [CrossRef]
58. Johansson, F.; Brodin, T. Effects of Fish Predators and Abiotic Factors on Dragonfly Community Structure. *J. Freshw. Ecol.* **2003**, *18*, 415–423. [CrossRef]
59. Fuiman, L.A.; Magurran, A.E. Development of predator defences in fishes. *Rev. Fish Biol. Fish.* **1994**, *4*, 145–183. [CrossRef]
60. Campos-Sánchez, J.C.; Esteban, M.Á. Review of inflammation in fish and value of the zebrafish model. *J. Fish Dis.* **2021**, *44*, 123–139. [CrossRef]
61. Kirsten, K.; Soares, S.M.; Koakoski, G.; Carlos Kreutz, L.; Barcellos, L.J.G. Characterization of sickness behavior in zebrafish. *Brain Behav. Immun.* **2018**, *73*, 596–602. [CrossRef]
62. Chen, Q.; Gundlach, M.; Yang, S.; Jiang, J.; Velki, M.; Yin, D.; Hollert, H. Quantitative investigation of the mechanisms of microplastics and nanoplastics toward zebrafish larvae locomotor activity. *Sci. Total Environ.* **2017**, *584–585*, 1022–1031. [CrossRef]
63. Amran, N.H.; Zaid, S.S.M.; Mokhtar, M.H.; Manaf, L.A.; Othman, S. Exposure to Microplastics during Early Developmental Stage: Review of Current Evidence. *Toxics* **2022**, *10*, 597. [CrossRef]
64. Jeong, S.; Jang, S.; Kim, S.S.; Bae, M.A.; Shin, J.; Lee, K.-B.; Kim, K.-T. Size-dependent seizurogenic effect of polystyrene microplastics in zebrafish embryos. *J. Hazard. Mater.* **2022**, *439*, 129616. [CrossRef] [PubMed]
65. Li, X.; Li, X.; Chen, D.; Guo, J.-L.; Feng, D.-F.; Sun, M.-Z.; Lu, Y.; Chen, D.-Y.; Zhao, X.; Feng, X.-Z. Evaluating the biological impact of polyhydroxyalkanoates (PHAs) on developmental and exploratory profile of zebrafish larvae. *RSC Adv.* **2016**, *6*, 37018–37030. [CrossRef]
66. Lee, C.-W.; Hsu, L.-F.; Wu, I.L.; Wang, Y.-L.; Chen, W.-C.; Liu, Y.-J.; Yang, L.-T.; Tan, C.-L.; Luo, Y.-H.; Wang, C.-C.; et al. Exposure to polystyrene microplastics impairs hippocampus-dependent learning and memory in mice. *J. Hazard. Mater.* **2022**, *430*, 128431. [CrossRef] [PubMed]
67. Laue, K.; Jänicke, M.; Plaster, N.; Sonntag, C.; Hammerschmidt, M. Restriction of retinoic acid activity by Cyp26b1 is required for proper timing and patterning of osteogenesis during zebrafish development. *Development* **2008**, *135*, 3775–3787. [CrossRef]
68. Addison, M.; Xu, Q.; Cayuso, J.; Wilkinson, D.G. Cell Identity Switching Regulated by Retinoic Acid Signaling Maintains Homogeneous Segments in the Hindbrain. *Dev. Cell* **2018**, *45*, 606–620.e603. [CrossRef]
69. Kögel, T.; Bjorøy, Ø.; Toto, B.; Bienfait, A.M.; Sanden, M. Micro- and nanoplastic toxicity on aquatic life: Determining factors. *Sci. Total Environ.* **2020**, *709*, 136050. [CrossRef]

toxics

MDPI

Article

Microplastic Quantification in Aquatic Birds: Biomonitoring the Environmental Health of the Panjkora River Freshwater Ecosystem in Pakistan

Muhammad Bilal [1], Atif Yaqub [1], Habib Ul Hassan [2,3,*], Sohail Akhtar [4], Naseem Rafiq [5], Muhammad Ishaq Ali Shah [6], Ibrar Hussain [7], Muhammad Salman Khan [5], Asad Nawaz [8], Salim Manoharadas [9], Mohammad Rizwan Khan [10], Takaomi Arai [11] and Patricio De Los Ríos-Escalante [12]

1 Department of Zoology, Government College University Lahore, Lahore 54000, Pakistan; m.bilaldigaan@gmail.com (M.B.); atif@gcu.edu.pk (A.Y.)
2 Department of Zoology, University of Karachi, Karachi 75270, Pakistan
3 Fisheries Development Board, Ministry of National Food Security and Research, Islamabad 44000, Pakistan
4 Department of Mathematics and Statistics, University of Haripur, Haripur 22620, Pakistan; s.akhtar@uoh.edu.pk
5 Department of Zoology, Abdul Wali Khan University, Mardan 23200, Pakistan; naseem@awkum.edu.pk (N.R.); salman.khan@awkum.edu.pk (M.S.K.)
6 Department of Chemistry, Abdul Wali Khan University, Mardan 23200, Pakistan
7 Department of Statistics, Government College University Lahore, Lahore 54000, Pakistan
8 Institute for Advanced Study, Shenzhen University, Shenzhen 518060, China; 007298@yzu.edu.cn
9 Department of Botany and Microbiology, College of Science, King Saud University, Riyadh 11451, Saudi Arabia; smanoharadas@ksu.edu.sa
10 Department of Chemistry, College of Science, King Saud University, Riyadh 11451, Saudi Arabia; mrkhan@ksu.edu.sa
11 Environmental and Life Sciences Programme, Faculty of Science, Universiti Brunei Darussalam, Gadong BE 1410, Brunei; takaomi.arai@ubd.edu.bn
12 Facultad de Recursos Naturales, Departamento de Ciencias Biológicas y Químicas, Universidad Católica de Temuco, Temuco 4780000, Chile; prios@uct.cl
* Correspondence: habib5447@gmail.com or habib.ulhassan@yahoo.com; Tel.: +92-3455959739

Abstract: Microplastic pollution has become a global concern, with potential negative impacts on various ecosystems and wildlife species. Among these species, ducks (*Anas platyrhynchos*) are particularly vulnerable due to their feeding habits and proximity to aquatic environments contaminated with microplastics. The current study was designed to monitor microplastic (MP) pollutants in the freshwater ecosystem of the Panjkora River, Lower Dir, Pakistan. A total of twenty (20) duck samples were brought up for four months and 13 days on the banks of the river, with no food intake outside the river. When they reached an average weight of 2.41 $\pm$ 0.53 kg, all samples were sacrificed, dissected, and transported in an ice box to the laboratory for further analysis. After sample preparation, such as digestion with 10% potassium hydroxide (KOH), density separation, filtration, and identification, the MP content was counted. A total of 2033 MP particles were recovered from 20 ducks with a mean value of 44.6 $\pm$ 15.8 MPs/crop and 57.05 $\pm$ 18.7 MPs/gizzard. MPs detected in surface water were 31.2 $\pm$ 15.5 MPs/L. The major shape types of MPs recovered were fragments in crop (67%) and gizzard (58%) samples and fibers in surface water (56%). Other types of particles recovered were fibers, sheets, and foams. The majority of these detected MP particles were in the size range of 300–500 μm (63%) in crops, and 50–150 μm (55%) in gizzards, while in water samples the most detected particles were in the range of 150–300 μm (61%). Chemical characterization by FTIR found six types of polymers. Low-density polyethylene (LDPE) had the greatest polymer detection rate (39.2%), followed by polyvinyl chloride (PVC) (28.3%), high-density polyethylene (HDPE) (22.7%), polystyrene (6.6%), co-polymerized polypropylene (2.5%), and polypropylene homopolymer (0.7%). This study investigated the presence of microplastics in the crops and gizzards of ducks, as well as in river surface water. The results revealed the significant and pervasive occurrence of microplastics in both the avian digestive systems and the surrounding water environment. These findings highlight the potential threat of microplastic pollution to wildlife and ecosystems, emphasizing the need for further research and effective mitigation strategies to address this pressing environmental concern.

Keywords: microplastic pollution; freshwater; aquatic birds; Panjkora River; Pakistan

1. Introduction

Over the past few decades, the ubiquity and persistence of microplastic pollution have become a growing concern worldwide [1,2]. Global plastic production has increased the threat to society by contaminating the environment [3]. Particularly in aquatic environments, microplastics are contributing to excessive water pollution, and the ingestion of these particles is suspected to be harmful to organisms [1,3]. Microplastics, defined as particles smaller than 5 mm in size, have emerged as a significant environmental threat due to their abundance and potential to accumulate in various ecosystems [4,5]. Microplastics can enter freshwater environments in several ways, mostly via float-up garbage and waste material, but also through shared drain overflows, and degraded plastic debris from industrial effluent [6]. The issue of microplastic pollution gets worse as plastic particles break down into smaller particles as a result of various physical, chemical, and environmental factors [2–4]. Based on the sources, microplastics are divided into primary and secondary categories [3,7]. Small pieces of plastic called primary microplastics can enter the environment directly or indirectly through an overspill. Spills, sewage, and domestic and industrial effluent are examples of direct emissions of primary microplastics [8]. Secondary microplastics are formed from larger plastic items present in the environment when these larger plastic items disintegrate into smaller particles due to different physical, chemical, and environmental factors [6,9]. The amount of microplastic pollution is growing every day, increasing the risk of ecosystem exposure, and it is estimated that eight million tons of polyethylene bags leak into our aquatic ecosystem every year [6]. If plastics are produced and managed at the current rate, 12 billion tons of plastic waste will be discarded in landfills or the natural environment by 2050 [10]. Once in the environment, abiotic and biotic processes involving chemical, physical, and biological reactions can lead to the degradation of plastic waste at a very slow rate, generating numerous smaller plastics—and plastic manufacturing is expected to double [10]. Because of this increased risk of particle interaction with organisms, ingestion, adsorption, physical entanglement, and dangerous impacts across food webs, microplastics have the potential to have a wide range of effects on biota [3,11,12].

Plastic waste and debris have caused substantial environmental pollution globally in recent decades, and they have accumulated in hundreds of terrestrial and aquatic avian species. Birds are susceptible and vulnerable to external environments; therefore, they could be used to estimate the negative effects of environmental pollution [6,11]. Aquatic birds are especially susceptible to the negative impacts of microplastics because of their dependence on marine and freshwater ecosystems [6,13]. They perform important ecological roles as markers of environmental health and biodiversity since they are a diversified collection of avian species that live in coastal and inland environments [14]. The uptake of microplastics by aquatic birds is a sign of the intricate relationships that exist between avian species and their surroundings. These minute synthetic polymer pieces, which come from plastic bottles, fishing nets, and microbeads, are small enough to be mistaken for food [15]. Birds unintentionally swallow these particles with their prey or directly from contaminated water as they hunt in water bodies teeming with microplastics. As a result, microplastics may build up inside the bodies of the birds, which may have a variety of negative physiological, behavioral, and ecological effects [3,13,14].

Numerous studies have shown that microplastics (MPs) are ingested by aquatic, terrestrial, and avian species and retained in different sections of the gastrointestinal tract [10,16,17]. Ingestion and retention can be linked to a variety of physical, physiological, neural, and hormonal issues, chiefly a reduction in the area of the intestine that is used to absorb nutrients, inhibition of gastric and pancreatic enzymatic activity, decreased steroid hormones, delayed ovulation, cytotoxicity, lipid oxidative damage in gills and muscle, and neurotoxicity through lipid oxidative damage [3,5,10,11,13]. Microplastics can act as carriers for toxic chemicals, and contain many organic and inorganic contaminants, including heavy metal polychlorinated biphenyl (PCB) pollutants, the hazardous impacts of which are important to consider because these MPs further increase the likelihood of toxicity [5,7,11,18]. MPs absorb and excrete heavy metals within living organisms, for instance in the digestive tract, where desorption is facilitated by a low-pH environment [3]. The ingestion of microplastics by aquatic birds is a prevalent concern due to their feeding habits and the high abundance of microplastics in their foraging environments [14]. In wildlife, there are increasing reports of the ingestion of plastics across a wide range of taxa, including birds, with detrimental effects on health [7,10]. These effects include physical impairments such as intestinal blockage, ulcers, and perforation of the gut, and fake satiety as well as toxicological effects such as reproductive disorders, activation of inflammatory responses, and immunodeficiency, which might lead to increased mortality [4,10]. A study by Neves [19] found that 50% of ingested prey items in northern gannets (*Morus bassanus*) from the North Sea contained microplastics. Similarly, a study by Provencher [20] reported the presence of microplastics in the gastrointestinal tracts of multiple species of seabirds, including northern fulmars (*Fulmarus glacialis*) and black-legged kittiwakes (*Rissa tridactyla*). Microplastics can have detrimental physiological effects on aquatic birds, primarily through mechanical obstruction, inflammation, and chemical toxicity. A study by Ziccardi [21] demonstrated that ingestion of microplastics resulted in gut obstruction, reduced feeding efficiency, and altered nutrient absorption in common murres (*Uria aalge*). Additionally, Rummel [22] found that microplastics induce oxidative stress, inflammation, and tissue damage in several seabird species, including great shearwaters (*Ardenna gravis*) and Cory's shearwaters (*Calonectris diomedea*). A study by Bour [23] observed that microplastics in greater scaup (*Aythya marila*) contained higher concentrations of persistent organic pollutants compared to the surrounding sediments, indicating their potential to transfer chemicals to higher trophic levels. The ingestion of microplastics has been linked to immunotoxic effects in aquatic birds. Bond [24] found that exposure to microplastics resulted in suppressed immune responses in European herring gulls (*Larus argentatus*). Additionally, it caused endocrine disruption, fertility impairments, and reduced reproductive success in European starlings (*Sturnus vulgaris*).

Recently, studies have been published across the globe to report and highlight the MPs issue in birds, especially in aquatic ecosystems [6,11,13,14,25–27]. However, there has been no such attempt from Pakistan to highlight this concerning issue. Understanding the severity of the impact of microplastic pollution on aquatic birds necessitates a comprehensive assessment of their ecological significance. The present study aims to contribute to the growing body of knowledge regarding microplastic pollution in aquatic birds. By addressing the sources, effects, and implications of microplastics on avian health and their ecosystems, we hope to shed light on this emerging environmental issue and stimulate further research towards effective conservation strategies.

2. Material and Methods

2.1. Duckling Rearing and Sacrifice

A total of twenty (20) duckling (*Anas platyrhynchos*) samples were hatched and brought up for four months and 13 days on the banks of Panjkora River (34.768449, 71.792282), Lower Dir, KP, Pakistan. The total food intake of the ducklings was from the river. When they reached an average weight of 2.41 $\pm$ 0.53 kg, all samples were sacrificed and dissected, having had no food intake outside the river, and the crops and gizzards were transferred to beakers, labeled, and stored in the freezer for further analysis (Figure 1). Surface water samples were collected with a 1 L glass jar. The lid was removed, and then the glass jar was dipped into the water and samples were taken just one inch below the surface. When the jar was filled, it was recapped and stored in an ice box [7].

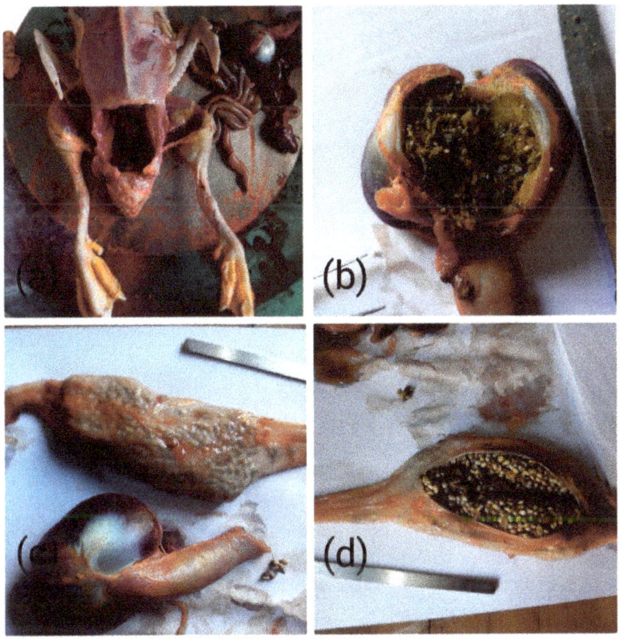

Figure 1. Dissection of duck: (**a**) Open gizzard with semi-digested content. (**b**) Closed crop and gizzard. (**c**) Open crop with ingested content (**d**).

Sample Preparation (Digestion, Density Separation, and Filtration)

Potassium hydroxide 10% (KOH) was added to a beaker containing crop and gizzard at a ratio of 5:1 to digest the sample. Then, it was stored in a water bath for 36 h at 55 °C. KOH was suggested for digestion since it is believed to have little effect on the decomposition of microplastics during the digestive process [3]. Density separation was used to separate the microplastic components. Sodium chloride (3:1 v/v) was added after digestion, and the mixture was agitated for 20 min before settling for 24 h [12]. After a 24 h period of settlement, the sample's supernatant layer was removed, and three distinct sizes of fractions were obtained by passing each through sieves with varying pore sizes (500 µm, 300 µm, 150 µm, and 50 µm). Each fraction was then filtered using filter paper in a filtering assembly. After filtration, the filter assembly cup's walls were cleaned twice, and the solid-containing filter paper was left in a Petri dish to dry for a day before being ready for detection [3,12]. The same procedure was followed for water samples except that in the digestion step, Fenton's reagent, an acidic solution of ferrous sulfate, and hydrogen peroxide (35%) were added to 200 mL of the sample, respectively, to digest any organic compounds that might have been present.

2.2. *Microplastics Observation, Identification, and Quantification*

A light binocular microscope (at 16 × 4 and 16 × 10 magnification, Labomed, model: CXL-110446002, 9135002, New York, NY, USA) was used for observation and inspections of dried filter papers containing particles. All particles of MPs were counted manually under microscopic observation. Images of the discovered particles were taken using a 1600× USB 8 LEDs electronic digital microscope camera and a Zeiss stereomicroscope Stemi 508 microscope at 2.5 magnification. In the current study, we did not measure any particles smaller than 50 µm. To identify categories, physical attributes including size, geometry, and color were taken into account. The polymer spectrum library of Omnic Spectra (Version 7.3, Thermo Fisher Scientific Inc., Waltham, MA, USA) software was used to perform polymer identification FTIR spectroscopy (IRTracer-100, Shimadzu, Columbia, MD, USA).

2.3. Background Contamination Control and Limitations of The Study

To keep samples from being contaminated by air, every safety precaution was taken. When not in use, all laboratory tools, including glassware and chemicals, were covered with aluminum foil. To prevent environmental contamination, distilled water, reagents, and other materials were filtered and covered in aluminum foil. To determine the suspended load of MPs from the environment, a few filter papers were scattered throughout the lab for 72 h in various locations. These filter sheets were then viewed with a stereomicroscope. Six of these filter sheets were evaluated and preserved as a control for the analysis. Any particle less than 50 µm was not considered in this study due to the non-availability of equipment for fine detection.

3. Results and Discussion

A total of 2033 MP particles were recovered from 20 duck samples combined from crops and gizzards, where 892 MP particles were recovered from 20 crop samples of ducks with a mean value of 44.6 $\pm$ 15.8 MPs/crop, and 1141 MPs from gizzard samples with a mean value of 57.05 $\pm$ 18.7 MPs/gizzard. Meanwhile, a total of 625 MP particles were detected from 20 samples of surface water of the river in which these ducks were reared. The mean of MPs detected in surface water was 31.2 $\pm$ 15.5 MPs/L (Figures 2 and 3). A weak correlation was seen to exist between the concentration of MPs in river water and the crop of a duck ($r = 0.24$), while a very weak correlation was observed between the concentration of MPs in river water and gizzards ($r = 0.058$) (Table 1). The possible potential reason for the weak correlation may be the loss of MP particles from crops and gizzards through the gastrointestinal canal in feces. Various studies across the globe have highlighted the MPs issue in aquatic birds; for example, Susanti et al. [28] collected 25 duck samples for the assessment of MPs. They found 27 to 41 MPs/individual in the gastrointestinal tracts of the ducks. Another study, by Bustamante [29], assessed and found MPs in the gizzards of Virginia waterfowl. In his study, the author assessed some species such as mallard (Anas platyrhynchos), long-tailed duck (Clangula hyemalis), goldeneye duck (Bucephala clangula), Canada goose (Branta canadensis), and ringneck duck (Aythya collaris). He recovered MP particles from 53.6% of the gizzards of waterfowls. The abundance range he found was 0 to 1.75 MPs/gram of gizzard material. Faure et al. [30] evaluated and carried out sampling in Lake Geneva, gathering samples from the following species: *Cygnus olor* (Gmelin, 1789), *Anas platyrhynchos* (L., 1758), and Ardea cinerea (Linnaeus, 1758); MPs were found in the gastrointestinal (GI) tracts of eight of the nine birds. When further *Cinclus cinclus* (Linnaeus, 1758) specimens were taken, regurgitates and fecal samples were used to determine the prevalence of MPs, which was discovered to be 50% and 45%, respectively [31]. A broader study investigated the MP ingestion in 350 samples from 17 species, including a marine one. It revealed an anthropogenic debris ingestion rate of 11.1%. According to an extrapolation of the findings limited to plastic, 9.7% of freshwater species include MPs [32]. In addition to research on adult birds, Laurentian Great Lakes' *Phalacrocorax auritus* (Lesson, 1831) chicks were dissected. According to Brookson et al. [33], the majority of MP fibers were found in the gastrointestinal (GI) tracts of over 86% of the chicks. Numerous studies have examined how marine debris is consumed by seabirds [34] and microplastics, which are essentially pellets and user fragments, have been isolated from cadavers, regurgitated samples, and feces of birds used in the studies [35–38]. Seabirds may be able to regurgitate microplastics from their digestive tracts after consumption [39]. On the other hand, this implies that parents might expose their chicks to plastic while feeding them. This is corroborated by Kühn and van Franeker's [40] discovery that juvenile intestines contain more plastic than adult ones. This may suggest that the majority of bird microplastic contamination occurs between generations and that the act of regurgitation may cause the degradation of microplastics into even smaller particles. The majority of the birds investigated did not pass away as a direct result of ingesting microplastic, hence it may be deduced that seabirds are not as seriously harmed by microplastic ingestion as they are by macroplastic ingestion [41]. There is currently no evidence that microplastics can cross the intestine barrier, enter the bloodstream, or accumulate in various organs because the majority of studies on microplastics in seabirds only examined microplastics in the digestive tract and feces [42]. No research has shown nanometer-sized microplastics in the excrement or intestines of seabirds as of yet. (Table 2).

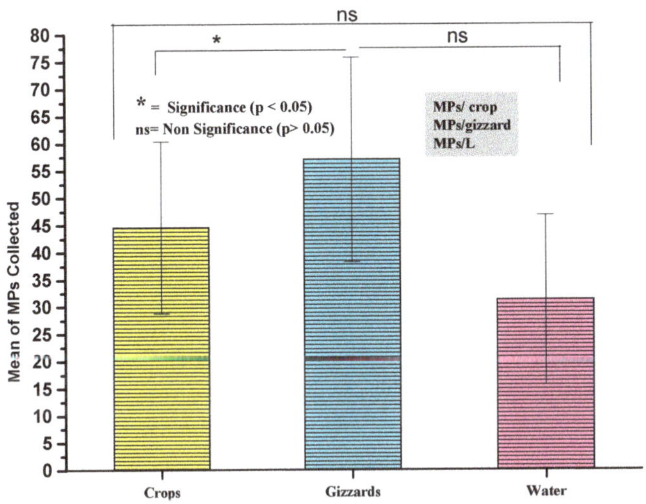

Figure 2. The mean concentration of MPs detected in crops, gizzards, and river water.

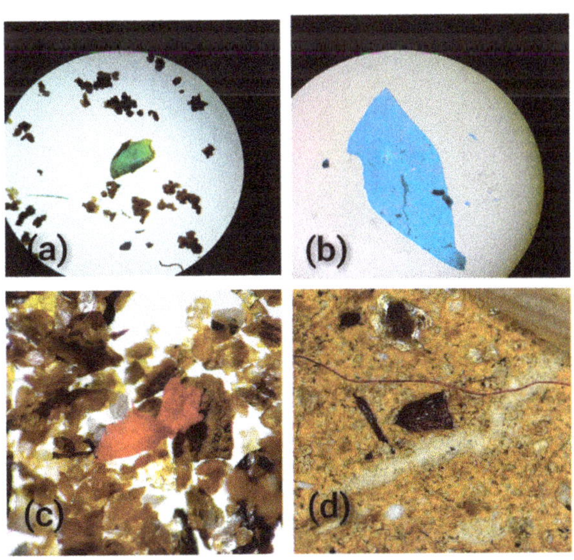

Figure 3. Microscopic images of MP fragment (**a**), sheet (**b**), foam (**c**), and fiber (**d**).

Table 1. Correlation between MP concentration in river water and the crops and gizzards of the ducks.

Correlations				
		Crops	Gizzards	Water
Water	Pearson Correlation	0.242	0.058	1
	Sig.	0.303	0.809	
	N	20	20	20

3.1. Shapes of Detected MPs

The most common shape types of MPs recovered were fragments in crop and gizzard samples and fibers in surface water. Of the total MPs detected in crops, 67% were fragments, while the other types of particles recovered from crops were fibers, sheets, and foams, with percentages of 23%, 8%, and 2%, respectively. Fragments were also the dominant type of MPs in gizzards as 58% of the detected MPs were fragments in gizzards. Other types of MPs found in gizzard samples were sheets (21%), fibers (18%), and foams (3%). Surprisingly, the dominant types of MPs detected in water samples were fibers with an abundance percentage of 56%. Fragments were the second most abundant type of particle in water samples, at 31%, while sheets and foams were 12% and 1%, respectively (Figures 3 and 4, Table 2). The majority of microplastic particles detected in avian gizzards and crops are fragments. For example, Collard [43] observed that 72.9% of the particles were fragments. Tokunaga [44] identified fragments as the predominant kind of microplastics in their investigation of wild birds in Japan. Zhao [45] found that fragments were a crucial sort of particle shape for the retrieved microplastics. In their research, 54.9% of MPs were fragments, while 37.4% were fibers. Unlike the current study, Deoniziak [46] indicated that fragments made up only 10% of the total particles and that fibers were predominant (84%) among the particles. Less mobility in the GIT tract and difficulty excreting through feces may be contributing factors to the preponderance of fragment-type particles in crops and gizzards. Another scenario is that the plastic pieces fool the birds into thinking they are food, and they eat them selectively (Table 2).

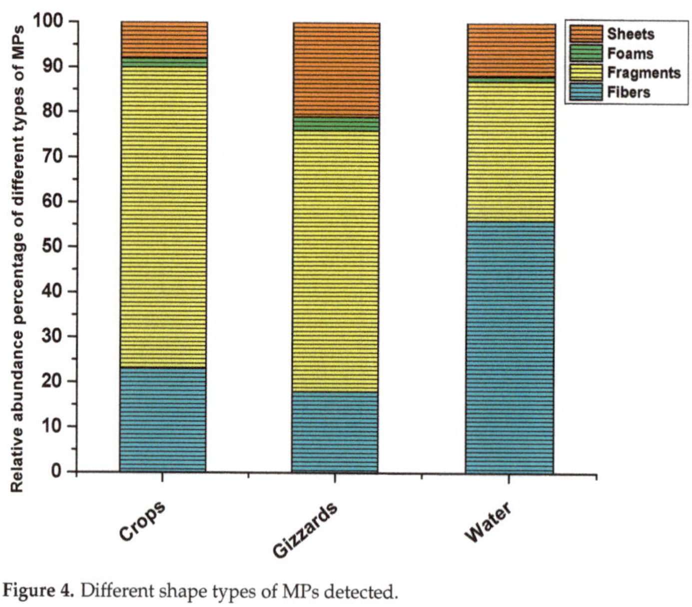

Figure 4. Different shape types of MPs detected.

3.2. Sizes of Detected MPs

In terms of size range, 63% of these detected MP particles were in the size range of 500–300 μm in crops, while the abundances of other size ranges of detected particles in crops were 27% (150–300 μm) and 10% (50–150 μm). Unlike in crops, the dominant size range of detected MPs in gizzards was 150–50 μm, at 55%. The abundance percentages of other size ranges in gizzards were 34% (300–500 μm) and 11% (150–300 μm). Meanwhile, in water samples, the most detected particles were in the range of 150–300 μm (61%). Other size ranges such as 300–500 μm and 50–150 μm were 15% and 24%, respectively (Figure 5 and Table 2). Various investigations conducted around the world have shown high concentrations of comparably bigger particles. Bessa et al. [47] retrieved 19 microplastic particles from the scat of penguins where the majority of the discovered particles were greater than 500 μm, and the mean size of the particles was 1266 ± 1378 μm. Liu [48] reported the presence of microplastics, with the majority of the particles being in the 500–1000 μm size range. Particles from the gastrointestinal (GI) tracts of birds were collected by Zhu [49] and included 92.9% of particles smaller than 5 mm, and more than 90% in a study by Zhao [45], while Deoniziak [46] gathered bird GIT particles that were smaller than 1000 μm. The majority (68.7%) of little-black cormorants' (Phalacrocorax sulcirostris) GIT-extracted particles were in the 100–1000 μm size range [50]. The fact that larger particles are less mobile in the GIT tract and consequently get stuck in different regions of the GIT, such as the stomach and gizzards, may be the cause of the significantly larger particle presence. The smaller particles, however, tend to pass through feces and are more easily moved by the GIT (Table 2).

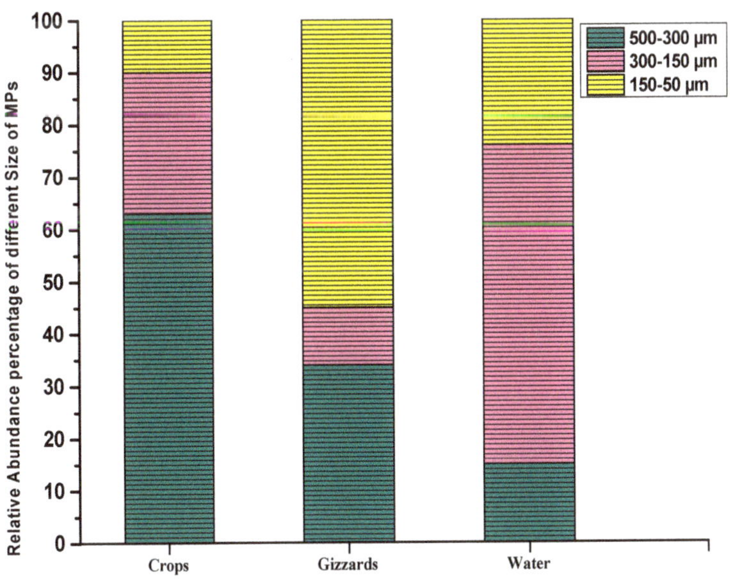

Figure 5. Size ranges of the detected MPs.

3.3. Detected Polymer Types of MPs

It was feasible to establish the chemical composition of the type of polymer by employing FTIR spectroscopy [51]. The absorbance peaks were noted after measuring the MP particles with an ATR sensor. The peak similarity index was additionally employed to evaluate the particle composition by contrasting recorded and reference peaks. Low-density polyethylene (LDPE), polyvinyl chloride (PVC), high-density polyethylene (HDPE), polystyrene (PS), co-polymer polypropylene (COPP), and polypropylene homopolymer (PPH) were the six types of polymers found. Low-density polyethylene (LDPE) had the greatest detection rate (39.2%), followed by polyvinyl chloride (PVC) (28.3%), high-density polyethylene (HDPE) (22.7%), polystyrene (6.6%), co-polymerized polypropylene (2.5%), and polypropylene homopolymer (0.7%) (Figure 6 and Table 2). While several studies have shown various polymer types, there was considerable global overlap in the polymer types found in microplastic particles. The majority of the microplastic particles, according to Collard [52], were of the polypropylene, polystyrene, and polyethylene variety. Polyethylene was identified as the most prevalent form of polymer. In a different study, polyethylene terephthalate (16%), ethylene-co-polypropylene (11%), and cellulose were shown to be more prevalent than the previous two types of polymers, making up 37% of the total particles [26]. In their research, Bessa et al. [47] discovered several polymers, including polypropylene, polyethylene, polyacrylonitrile, and polyacrylate. Other studies [51] have also referred to these polymers as being among the primary types of MPs. The majority of packaging employs LDPE and HDPE polymers, including foils, milk, shampoo, oil, and soap bottles, domestic items like trays, plates, and cups, cables, and PVC in electrical and electronic equipment, tour tents, and water pipes [9]. Packaging for a variety of items, such as structural tanks, battery covers, and pump components, has been made with PPH [53]. All of these plastic items, when thrown into or close to water bodies, shatter into small pieces and finally lead to the production of MPs with different polymer natures. In the current study, the major fraction of the load of these polymer types of MPs comes from wastewater discharge into the river from the local market containing factories of plastic pipes, spices, plastic shoes, etc., and also tourists frequently leaving behind disposable plates, glasses, water bottles, food wrappers, and other items near and in the river. It was concluded as a result that the identified polymers in this study were related to their potential application in the region under study (Table 2).

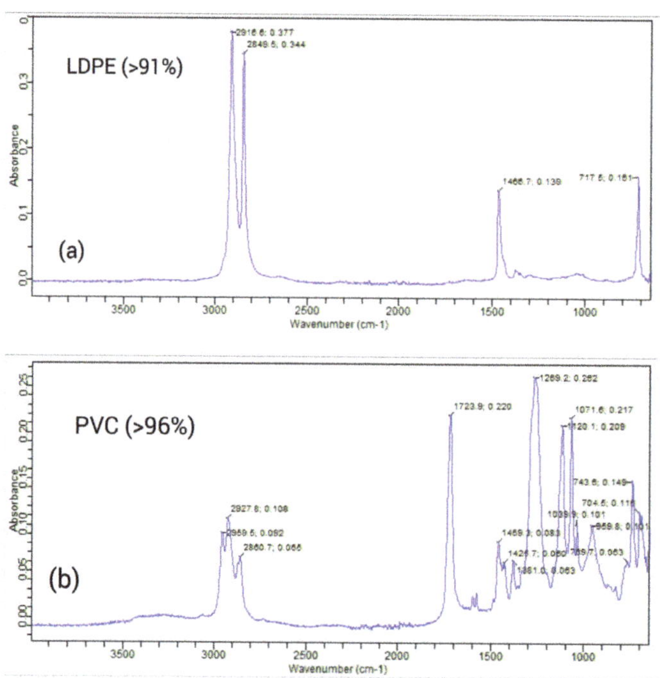

Figure 6. The top two most abundant types of polymers: LDPE (**a**) and PVC (**b**).

Table 2. Comparison of the present study with worldwide reports.

Particles	Region	Detected Particles	Reference
	Pakistan	MPs found with a mean of 44.6 ± 15.8 MPs/crop and 57.05 ± 18.7 MPs/gizzard of duck	Present Study
	Pakistan	Extracted 33.25 ± 17.8 MPs/gizzard, 17.8 ±12.1 MPs/crop of bird	[12]
	Indonesia	Found 27 to 41 MPs/duck	[28]
	Virginia, USA	Found 0 to 1.75 MPs/gram of gizzard of Virginia fowl	[29]
	Zurich and Brienz, Switzerland	MPs were present in eight of the nine birds	[30]
	South Wales, UK	MPs were found in 50% of regurgitates (n = 72) of Eurasian dipper (*Cinclus cinclus*)	[31]
	Canada	MPs found in 9.7% of freshwater species	[32]
	North America	MP fibers were found in the GI tracts of over 86% of the chicks of diving birds	[33]
	Norway	15–106 MPs from northern fulmar (*Fulmarus glacialis*)	[37]
	Antarctic regions (Bird Island, South Georgia and Signy Island, South Orkney Islands)	Retrieved 19 microplastic particles from the scat of penguins	[47]
Size	Region	Dominant size range detected	Reference
	Pakistan	The majority (63%) of the detected MP particles were in the size range of 500–300 μm in crops; the dominant (55%) size range of detected MPs in gizzard was 150–50 μm	Present Study
	Pakistan	300–500 μm	[12]
	China	More than 90% of particles were smaller than 5 mm	[45]
		Particles from bird GIT were smaller than 1000 μm	[46]
	Antarctic regions (Bird Island, South Georgia and Signy Island, South Orkney Islands)	The majority of the discovered particles from the scat of penguins were greater than 500 μm	[47]
	China	The majority of the particles were in the 500–1000 μm size range	[48]
	China	Of the total particles from the GIT of the birds collected, 92.9% were particles smaller than 5 mm	[49]
	Indonesia	The majority (68.7%) of the little-black cormorant's (Phalacrocorax sulcirostris) GIT-extracted particles were in the 100–1000 μm size range	[50]
Shape	Region	Dominant shape type detected	Reference
	Pakistan	67% of total detected MPs from crops were fragments while 58% of detected MPs were fragments in gizzards	Present Study
	Pakistan	Fragments (64%) in gizzards Fragments (53%) in crops	[12]
		The majority (72.9%) of microplastic particles detected in avian gizzards and crops were fragments.	[43]
	Japan	Fragments as the predominant kind of microplastics detected in their investigation of wild birds in Japan	[44]
	China	In this study, 54.9% were fragments, while 37.4% were fibers.	[45]
	Poland	Fragments made up only 10% of the total particles and fiber was predominant (84%) among the particles in common blackbirds (*Turdus merula*) and song thrushes (*Turdus philomelos*).	[46]

Table 2. *Cont.*

Particles	Region	Detected Particles	Reference
Polymer	Region	Polymer types detected	Reference
		Low-density polyethylene (LDPE) had the greatest polymer detection rate (39.2%), followed by polyvinyl chloride (PVC) (28.3%), and high-density polyethylene (HDPE) (22.7%); 6.6% of the material was polystyrene, 2.5% was co-polymerized polypropylene, and 0.7% was polypropylene Homopolymer	Present study
	Pakistan	(PVC) with 51.2%, low-density polyethylene (LDPE) (30.7%), polystyrene (PS) (13.6%), polypropylene homopolymer (PPH) (4.5%)	[12]
	Central Florida, USA	Polyethylene terephthalate (16%), ethylene-co-polypropylene (11%), and cellulose were shown to be more prevalent than the previous two types of polymers, making up 37% of the total particles	[26]
	Antarctic regions (Bird Island, South Georgia and Signy Island, South Orkney Islands)	Discovered several polymers, including polypropylene, polyethylene, polyacrylonitrile, and polyacrylate	[47]
	China	Polyethylene terephthalate (51%), epoxy resin (19%), polyethylene (12%), and alkyd resin (8%)	[48]
	Norway	Polypropylene, polystyrene, and polyethylene variety; polyethylene was identified as the most prevalent form of polymer	[52]

4. Conclusions

To our knowledge, this is the first study of MPs in the crops and gizzards of wild birds. The current study found MPs in all samples of crops and gizzards as well in river surface water. Overall, relatively higher numbers of MPs were recovered from gizzards than crops. Four different types and shapes of MP particles were recovered, and fragments were the dominant type in crops and gizzards, while in surface water, fibers were the most abundant type of MPs. In terms of the size range, relatively larger particles (300–500 μm) were abundant in crops, while the dominant size ranges in gizzards and water were 150–300 μm and 50–150 μm, respectively. Duck is a suitable species to assess aquatic plastic pollution. The current study could be a valuable bridge between already existing literature and future recommendations in this regard. As the issue of microplastics and their impact on the environment becomes more apparent, there is an urgent need for further research to address this problem effectively. This study provides new and important information on MP contamination in wild birds. Further research should be conducted to quantitively assess MP contamination in wild birds and to investigate the health implications associated with MP inhalation, particularly because MP pollution is expected to become more severe in the future on a global scale. Future research should include standardized methods, impact on human health, microplastic removal techniques, geographical variations, ecological factors, and other approaches to effectively address this crucial issue.

Author Contributions: Conceptualization, M.B. and H.U.H.; methodology, M.B. and H.U.H.; software and validation, A.Y., S.M. and M.R.K.; formal analysis N.R. and I.H.; investigation M.B. and H.U.H.; resources, M.I.A.S. and M.S.K.; data curation, S.M.; writing—original draft preparation and analysis, M.B. and H.U.H.; writing—review and editing A.N., T.A. and P.D.L.R.-E.; visualization, S.A.; project administration, M.B., A.Y. and H.U.H.; funding acquisition, P.D.L.R.-E. All authors have read and agreed to the published version of the manuscript.

Funding: This work was financially supported by the project MECESUP UCT 0804.

Institutional Review Board Statement: This study was carried out in strict accordance with the recommendations of the Local Ethical Committee for Experiments on Animals of the Government College University Lahore, Pakistan and UOK, Ref. No. ST/77/CEES Dated: 28 January 2020.

Informed Consent Statement: Not applicable. This manuscript does not contain data from any person.

Data Availability Statement: All data generated or analyzed during this study were included in this published article.

Acknowledgments: The authors thank the Research Supporting Project for funding this work through Research Supporting Project number (RSPD2023R708), King Saud University, Riyadh, Saudi Arabia.

Conflicts of Interest: The authors declare no conflict of interest.

References

1. Arowojolu, I.M.; de Oliveira Alves, I.; Benson, N.U.; Sodré, F.F. Microplastics in aquatic environments: A growing, unresolved concern. *Chem. Total Environ.* **2023**, *3*, 8–22.
2. Ferreira, H.C.; Lôbo-Hajdu, G. Microplastics in coastal and oceanic surface waters and their role as carriers of pollutants of emerging concern in marine organisms. *Mar. Environ. Res.* **2023**, *188*, 106021. [CrossRef] [PubMed]
3. Bilal, M.; Qadir, A.; Yaqub, A.; Hassan, H.U.; Irfan, M.; Aslam, M. Microplastics in water, sediments, and fish at Alpine River, originating from the Hindu Kush Mountain, Pakistan: Implications for conservation. *Environ. Sci. Pollut. Res.* **2022**, *30*, 727–738. [CrossRef] [PubMed]
4. de Oliveira, C.R.S.; da Silva Júnior, A.H.; Mulinari, J.; Ferreira, A.J.S.; da Silva, A. Fibrous microplastics released from textiles: Occurrence, fate, and remediation strategies. *J. Contam. Hydrol.* **2023**, *256*, 104169. [CrossRef]
5. Zitouni, N.; Cappello, T.; Missawi, O.; Boughattas, I.; De Marco, G.; Belbekhouche, S.; Mokni, M.; Alphonse, V.; Guerbej, H.; Bousserrhine, N.; et al. Metabolomic disorders unveil hepatotoxicity of environmental microplastics in wild fish Serranus scriba (Linnaeus 1758). *Sci. Total Environ.* **2022**, *10*, 155872. [CrossRef]
6. Yin, L.; Du, L.; Wen, X.; Huang, D.; Xiao, R.; Wang, Z.; Su, H.; Huang, J.; Wang, G.; Tao, J.; et al. Occurrence and Effects of Microplastics in Lake Ecosystems: Particular Focus on Migration in Water and Food Chains. *Rev. Environ. Cont. Toxicol.* **2023**, *261*, 11. [CrossRef]
7. Bilal, M.; Ul Hassan, H.; Siddique, M.A.M.; Khan, W.; Gabol, K.; Ullah, I.; Sultana, S.; Abdali, U.; Mahboob, S.; Khan, M.S.; et al. Microplastics in the Surface Water and Gastrointestinal Tract of Salmo trutta from the Mahodand Lake, Kalam Swat in Pakistan. *Toxics* **2023**, *11*, 3. [CrossRef]
8. Hanslik, L. Microplastics in Limnic Ecosystems-Investigation of Biological Fate and Effects of Microplastic Particles and Associated Contaminants in Zebrafish (Danio rerio). Ph.D. Thesis, Heidelberg University, Heidelberg, Germany, 2020.
9. Irfan, M.; Qadir, A.; Mumtaz, M.; Ahmad, S.R. An unintended challenge of microplastic pollution in the urban surface water system of Lahore, Pakistan. *Environ. Sci. Pollut. Res.* **2020**, *27*, 16718–16730. [CrossRef]
10. Bilal, M.; Ul Hassan, H.; Taj, M.; Rafiq, N.; Nabi, G.; Ali, A.; Gabol, K.; Shah, M.I.A.; Ghaffar, R.A.; Sohail, M.; et al. Biological Magnification of Microplastics: A Look at the Induced Reproductive Toxicity from Simple Invertebrates to Complex Vertebrates. *Water* **2023**, *15*, 2831. [CrossRef]
11. Dellisanti, W.; Leung, M.M.; Lam, K.W.; Wang, Y.; Hu, M.; Lo, H.S.; Fang, J.K. A short review on the recent method development for extraction and identification of microplastics in mussels and fish, two major groups of seafood. *Mar. Pollut. Bull.* **2023**, *186*, 114221. [CrossRef]
12. Bilal, M.; Taj, M.; Ul Hassan, H.; Yaqub, A.; Shah, M.I.A.; Sohail, M.; Rafiq, N.; Atique, U.; Abbas, M.; Sultana, S.; et al. First Report on Microplastics Quantification in Poultry Chicken and Potential Human Health Risks in Pakistan. *Toxics* **2023**, *11*, 612. [CrossRef] [PubMed]
13. Barboza, L.G.A.; Lopes, C.; Oliveira, P.; Bessa, F.; Otero, V.; Henriques, B.; Raimundo, J.; Caetano, M.; Vale, C.; Guilhermino, L. Microplastics in Wild Fish from North East Atlantic Ocean and Its Potential for Causing Neurotoxic Effects, Lipid Oxidative Damage, and Human Health Risks Associated with Ingestion Exposure. *Sci. Total Environ.* **2020**, *717*, 134625. [CrossRef] [PubMed]
14. Bange, A.; Backes, A.; Garthe, S.; Schwemmer, P. Prey choice and ingestion of microplastics by common shelducks and common eiders in the Wadden Sea World Heritage Site. *Mar. Biol.* **2023**, *170*, 54. [CrossRef]
15. Cole, M.; Lindeque, P.; Fileman, E.; Halsband, C.; Galloway, T.S. The impact of polystyrene microplastics on feeding, function and fecundity in the marine copepod *Calanus helgolandicus*. *Environ. Sci. Technol.* **2015**, *49*, 1130–1137. [CrossRef] [PubMed]
16. Khan, M.L.; Hassan, H.U.; Khan, F.U.; Ghaffar, R.A.; Rafiq, N.; Bilal, M.; Khooharo, A.R.; Ullah, S.; Jafari, H.; Nadeem, K.; et al. Effects of microplastics in freshwater fishes health and the implications for human health. *Braz. J. Biol.* **2023**, *84*, e272524. [CrossRef] [PubMed]
17. Bilal, M.; Ali, H.; Ullah, R.; Hussain, I.; Khan, B.A. Overview of Microplastics Threat in Aquatic Animals Since 1960 to 2020. *Int. J. Res. Anal. Rev.* **2021**, *8*, 950–964.
18. Porcino, N.; Bottari, T.; Mancuso, M. Is wild marine biota affected by microplastics? *Animals* **2022**, *13*, 147. [CrossRef]

19. Neves, D.; Sobral, P.; Ferreira, J.L.; Pereira, T. Ingestion of microplastics by commercial fish off the Portuguese coast. *Mar. Pollut. Bull.* **2015**, *101*, 119–126. [CrossRef]
20. Provencher, J.F.; Vermaire, J.C.; Avery-Gomm, S.; Braune, B.M.; Mallory, M.L. Garbage in guano? Microplastic debris found in faecal precursors of seabirds known to ingest plastics. *Sci. Total Environ.* **2018**, *644*, 1477–1484. [CrossRef]
21. Ziccardi, L.M.; Edgington, A.; Hentz, K.; Kulacki, K.J.; Kane Driscoll, S. Microplastics as vectors for bioaccumulation of hydrophobic organic chemicals in the marine environment: A state-of-the-science review. *Environ. Toxicol. Chem.* **2016**, *35*, 1667–1676. [CrossRef]
22. Rummel, C.D.; Escher, B.I.; Sandblom, O.; Plassmann, M.M.; Arp, H.P.H.; MacLeod, M.; Jahnke, A. Effects of leachates from UV-weathered microplastic in cell-based bioassays. *Environ. Sci. Technol.* **2019**, *53*, 9214–9223. [CrossRef] [PubMed]
23. Bour, A.; Avio, C.G.; Gorbi, S.; Regoli, F.; Hylland, K. Presence of microplastics in benthic and epibenthic organisms: Influence of habitat, feeding mode and trophic level. *Environ. Pollut.* **2018**, *243*, 1217–1225. [CrossRef] [PubMed]
24. Bond, A.L. Diet changes in breeding herring gulls (*Larus argentatus*) in Witless Bay, Newfoundland and Labrador, Canada, over 40 years. *Waterbirds* **2016**, *39*, 152–158. [CrossRef]
25. Masiá, P.; Ardura, A.; Garcia-Vazquez, E. Microplastics in special protected areas for migratory birds in the Bay of Biscay. *Mar. Pollut. Bull.* **2019**, *146*, 993–1001. [CrossRef]
26. Carlin, J.; Craig, C.; Little, S.; Donnelly, M.; Fox, D.; Zhai, L.; Walters, L. Microplastic accumulation in the gastrointestinal tracts in birds of prey in central Florida, USA. *Environ. Pollut.* **2020**, *264*, 114633. [CrossRef]
27. Wang, L.; Nabi, G.; Yin, L.; Wang, Y.; Li, S.; Hao, Z.; Li, D. Birds and plastic pollution: Recent advances. *Avian Res.* **2021**, *12*, 1–9. [CrossRef]
28. Susanti, R.; Yuniastuti, A.; Fibriana, F. The Evidence of microplastic contamination in Central Javanese local ducks from intensive animal husbandry. *Water Air Soil Pollut.* **2021**, *232*, 178. [CrossRef]
29. Bustamante, T. Assessing the Presence and Concentration of Microplastics in the Gizzards of Virginia Waterfowl. Ph.D. Thesis, University of Mary Washington, Washington, VA, USA, 2021.
30. Faure, F.; Corbaz, M.; Baecher, H.; de Alencastro, L.F. Pollution due to plastics and microplastics in Lake Geneva and in the Mediterranean Sea. *Arch. Des. Sci.* **2012**, *65*, 157–164.
31. D'Souza, J.M.; Windsor, F.M.; Santillo, D.; Ormerod, S.J. Food web transfer of plastics to an apex riverine predator. *Glob. Chang. Biol.* **2020**, *26*, 3846–3857. [CrossRef]
32. Holland, E.; Mallory, M.; Shutler, D. Plastics and other anthropogenic debris in freshwater birds from Canada. *Sci. Total Environ.* **2016**, *571*, 251–258. [CrossRef]
33. Brookson, C.B.; de Solla, S.R.; Fernie, K.J.; Cepeda, M.; Rochmana, C.M. Microplastics in the diet of nestling double-crested cormorants (*Phalacrocorax auritus*), an obligate piscivore in a freshwater ecosystem. *Can. J. Fish. Aquat. Sci.* **2019**, *76*, 2156–2163. [CrossRef]
34. Kühn, S.; Rebolledo, E.L.B.; van Franeker, J.A. Deleterious effects of litter on marine life. In *Marine Anthropogenic Litter*; Springer: Berlin/Heidelberg, Germany, 2015; pp. 75–116.
35. Bond, A.L.; Provencher, J.F.; Daoust, P.-Y.; Lucas, Z.N. Plastic ingestion by fulmars and shearwaters at Sable Island, Nova Scotia, Canada. *Mar. Pollut. Bull.* **2014**, *87*, 68–75. [PubMed]
36. Codina-García, M.; Militão, T.; Moreno, J.; González-Solís, J. Plastic debris in Mediterranean seabirds. *Mar. Pollut. Bull.* **2013**, *77*, 220–226.
37. Herzke, D.; Anker-Nilssen, T.; Nøst, T.H.; Götsch, A.; Christensen-Dalsgaard, S.; Langset, M.; Fangel, K.; Koelmans, A.A. Negligible Impact of Ingested Microplastics on Tissue Concentrations of Persistent Organic Pollutants in Northern Fulmars off Coastal Norway. *Environ. Sci. Technol.* **2016**, *50*, 1924–1933. [CrossRef]
38. Tanaka, K.; Takada, H.; Yamashita, R.; Mizukawa, K.; Fukuwaka, M.-A.; Watanuki, Y. Accumulation of plastic-derived chemicals in tissues of seabirds ingesting marine plastics. *Mar. Pollut. Bull.* **2013**, *69*, 219–222.
39. Lindborg, V.A.; Ledbetter, J.F.; Walat, J.M.; Moffett, C. Plastic consumption and diet of Glaucous-winged Gulls (*Larus glaucescens*). *Mar. Pollut. Bull.* **2012**, *64*, 2351–2356.
40. Kühn, S.; van Franeker, J.A. Plastic ingestion by the northern fulmar (*Fulmarus glacialis*) in Iceland. *Mar. Pollut. Bull.* **2012**, *64*, 1252–1254. [CrossRef]
41. Lusher, A. Microplastics in the marine environment: Distribution, interactions and effects. In *Marine Anthropogenic Litter*; Springer: Berlin/Heidelberg, Germany, 2015; pp. 245–307.
42. Reynolds, C.; Ryan, P.G. Micro-plastic ingestion by waterbirds from contaminated wetlands in South Africa. *Mar. Pollut. Bull.* **2018**, *126*, 330–333.
43. Collard, F.; Bangjord, G.; Herzke, D.; Gabrielsen, G.W. Plastic burdens in northern fulmars from Svalbard: Looking back 25 years. *Mar. Pollut. Bull.* **2022**, *185*, 114333.
44. Tokunaga, Y.; Okochi, H.; Tani, Y.; Niida, Y.; Tachibana, T.; Saigawa, K.; Katayama, K.; Moriguchi, S.; Kato, T.; Hayama, S.-I. Airborne Microplastics Detected in the Lungs of Wild Birds in Japan. *Chemosphere* **2022**, *321*, 4276259.
45. Zhao, S.; Zhu, L.; Li, D. Microscopic anthropogenic litter in terrestrial birds from Shanghai, China: Not only plastics but also natural fibers. *Sci. Total Environ.* **2016**, *550*, 1110–1115. [PubMed]
46. Deoniziak, K.; Cichowska, A.; Nied'zwiecki, S.; Pol, W. Thrushes (Aves: Passeriformes) as indicators of microplastic pollution interrestrial environments. *Sci. Total Environ.* **2022**, *853*, 158621.

47. Bessa, F.; Ratcliffe, N.; Otero, V.; Sobral, P.; Marques, J.C.; Waluda, C.M.; Trathan, P.N.; Xavier, J.C. Microplastics in gentoo penguins from the Antarctic region. *Sci. Rep.* **2019**, *9*, 14191.
48. Liu, S.L.; Jian, M.F.; Zhou, L.Y.; Li, W.H.; Wu, X.E.; Rao, D. Pollution characteristics of microplastics in migratory bird habitats located within Poyang Lake wetlands. *Huan Jing Ke Xue Huanjing Kexue* **2019**, *40*, 2639–2646. [PubMed]
49. Zhu, C.; Li, D.; Sun, Y.; Zheng, X.; Peng, X.; Zheng, K.; Hu, B.; Luo, X.; Mai, B. Plastic debris in marine birds from an island located in the South China Sea. *Mar. Pollut. Bull.* **2019**, *149*, 110566.
50. Susanti, N.K.Y.; Mardiastuti, A.; Hariyadi, S. Microplastics in Digestive System of Little-black cormorant (*Phalacrocorax sulcirostris*) in Pulau Rambut Sanctuary. In *IOP Conference Series: Earth and Environmental Science*; IOP Publishing: Bristol, UK, 2022; Volume 950, p. 012003.
51. Zhao, S.; Wang, T.; Zhu, L.; Xu, P.; Wang, X.; Gao, L.; Li, D. Analysis of suspended microplastics in the Changjiang Estuary:Implications for riverine plastic load to the ocean. *Water Res.* **2019**, *161*, 560–569.
52. Collard, F.; Husum, K.; Eppe, G.; Malherbe, C.; Hallanger, I.G.; Divine, D.V.; Gabrielsen, G.W. Anthropogenic particles in sediment from an Arctic fjord. *Sci. Total Environ.* **2021**, *772*, 145575.
53. Allahvaisi, S. Polypropylene in the industry of food packaging. In *Polypropylene*; Dogan, F., Ed.; IntechOpen: London, UK, 2012; pp. 3–22.

toxics

MDPI

Article

Distribution Characteristics of Microplastics in Surface Seawater off the Yangtze River Estuary Section and Analysis of Ecological Risk Assessment

Xiao Ji [1,2,†], Shuaishuai Yan [1,†], Yanlong He [1], Haisheng He [1] and Hanqi Liu [1,2,*]

1 East China Sea Ecological Center, MNR (Ministry of Natural Resources), Shanghai 201206, China; jixiao@ecs.mnr.gov.cn (X.J.); yanshuaishuai@ecs.mnr.gov.cn (S.Y.); ylhe@ecs.mnr.gov.cn (Y.H.); hehaisheng2018@163.com (H.H.)

2 Key Laboratory of Marine Ecological Monitoring and Restoration Technology, Ministry of Natural Resources, Shanghai 201206, China

* Correspondence: liuhanqi@ecs.mnr.gov.cn

† These authors contributed equally to this work.

Abstract: Microplastics are widespread in the oceans as a new type of pollutant. Due to the special geographical environment characteristics, the Yangtze River estuary region become hotspot for microplastics research. In 2017 and 2019, surface seawater microplastics samples were collected from five stations off the Yangtze River estuary during four seasons (spring, summer, autumn, and winter). The abundance and characteristics of microplastics in seawater were researched. The results showed that microplastics widely existed in surface seawater; the average abundance of microplastics in seawater was (0.17 ± 0.14) items/m^3 (0.00561 ± 0.00462) mg/m^3; and accounting for 80% of the total plastic debris, the abundance of microplastics was at moderately low levels compared to national and international studies. The particle size of most microplastics was between 1 mm to 2 mm, accounting for 36.1% of the total microplastics. The main shapes of microplastics were fiber, flake, and line, accounting for 39.5%, 28.4%, and 20.8%, respectively. Polypropylene, polyethylene terephthalate, and polyethylene were the main components of microplastics, accounting for 41.0%, 25.1%, and 24.9%, respectively. Yellow, green, black, and transparent were the most common colors, accounting for 21.9%, 19.6%, 16.5%, and 15.7%, respectively. This study shows that the spatial distribution of microplastics in the surface waters off the Yangtze River estuary shows a decreasing trend from nearshore to farshore due to the influence of land-based inputs, hydrodynamics, and human activities; the distribution of microplastics has obvious seasonal changes, and the level of microplastic pollution is higher in summer. The potential ecological risk of microplastics in the surface waters off the Yangtze River estuary is relatively small.

Keywords: microplastic; surface seawater; Yangtze River estuary

1. Introduction

The concept of microplastics was first proposed by Thompson [1] in 2004. Seawater microplastics are plastic debris including fibers, particles, and fragments less than 5 mm in length that are present in the marine environment and mainly formed through the degradation of large plastic fragments by mechanical and photo-oxidation pathways [2–4]. These include primary and secondary microplastics [5]. Because microplastics are chemically stable and can exist in the environment for hundreds to thousands of years [6], they have received increasing attention as a new class of environmental pollutants. Microplastic pollution has become a global environmental problem that is widespread in the terrestrial and marine environments. Traces of microplastics have been found from land [7] to ocean [8–10], from nearshore [11,12] to pelagic to deep sea [13], from plateau lakes [14] to polar regions [15,16], from *Aeromonas* sp. [17,18] to marine mammals [19,20], and then to food [21] and cosmetics [22], which are closely related to human exposure.

Seawater microplastics have become a research hotspot for environmental issues in recent years, and scholars both domestic and abroad have carried out extensive studies, most of which focused on nearshore estuaries, bays, and other areas with intensive human activities. Scholars in China have carried out relevant studies in the Yangtze River estuary, South China Sea, Bohai Bay, etc., and the basic abundance ranges from 0.045 to 8.91 items/m^3. Microplastics in the surface waters of different seas in other regions of the world [23] have abundance ranges from 0.19 to 7.68 items/m^3. However, there are still fewer studies related to seasonal abundance changes.

Studies have mainly focused on offshore areas and beach surveys [24–30]. Therefore, in this paper, the seasonal distribution and compositional characteristics of microplastics in the outer surface seawater of the Yangtze River estuary were studied with the neighboring waters of the Yangtze River estuary as the study area to reveal the distribution characteristics of this microplastic and provide data support for the implementation of source control and management of microplastic pollution.

2. Materials and Methods

2.1. Study Area

The study area is located in the sea area off the Yangtze River estuary, with five stations (31.25°~32.42° N, 122.5°~124.5° E), and the section direction was laid out to coincide with the direction of the freshwater flushing in the Yangtze River estuary. Sampling was carried out in May 2017 (spring), November 2017 (autumn), February 2019 (winter), and August 2019 (summer), and the sampling stations are shown in Figure 1.

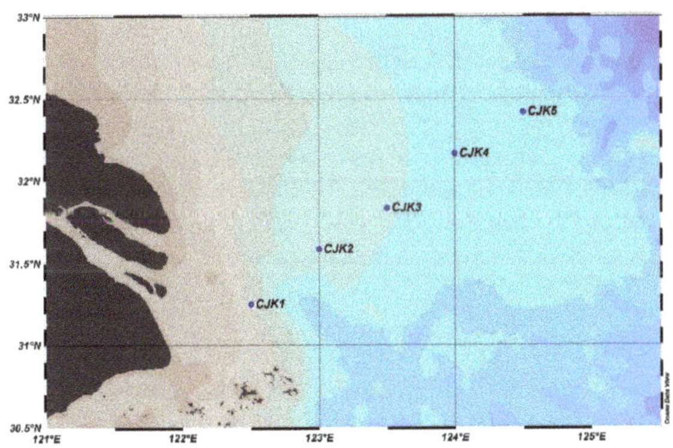

Figure 1. Sampling sites.

2.2. Sample Collection

In this study, a manta net was used to collect floating microplastics in surface seawater. The mouth of the net is 1 m long and 0.5 m wide; the net coat is a biological sampling net made of silk (the main component is protein), which is 3 m long, with an aperture of 330 μm, and a stainless-steel net bottom tube is connected at the bottom. There is a fixed flow meter in the center of the net mouth for calculating the amount of excess water. The vessel traveled at a speed of 2–3 knots, and each tow lasted 10–15 min. Before and after each tow, the net was rinsed, and the samples in the bottom tube were transferred to glass vials to avoid interfering with the samples between stations for further analysis in the laboratory.

2.3. Laboratory Analysis

The samples were passed through 5 mm and 330 μm stainless-steel mesh sieves sequentially, impurities such as fish and shrimp were rinsed and discarded, and the retained material from the 330 μm mesh sieve was rinsed with purified water and placed into a 500 mL beaker. Plastic samples of more than 5 mm were rinsed and stored separately for further analysis. The beakers and samples were dried in an oven at 60 °C. After drying, 20 mL of 0.05 mol/L ferrous sulfate solution and 20 mL of 30% hydrogen peroxide were added sequentially and digested at room temperature. If organic matter was still seen after digestion, we continued to add hydrogen peroxide and repeated the above operation until the organic matter in the sample was completely dissolved. After complete digestion, 6 g of NaCl solid was added to each 20 mL of digestion solution, dissolved, transferred to a glass funnel sealed at the lower end with a stopcock, and allowed to stand for density separation. The lower impurities were transferred to a beaker, and the supernatant was filtered using a glass fiber filter membrane (Whatman GF/F, 47 mm in diameter with a pore size of 0.7 μm), which was placed in a Petri dish and dried at 60 °C for further analysis.

The samples were observed using a stereo microscope to select the suspected plastic fragments, particles, etc., which were photographed using a stereo microscope with a photo-camera system (Nikon SMZ25, Nikon Corporation, Yokohama, Japan), and information on the physical characteristics of the plastics, such as shape, color, and size, was observed and documented using software (NIS-Elements D 4.50.00, Nikon Corporation, Yokohama, Japan) that was used in conjunction with the camera system. We used imageJ (1.5) software to first record the scale of the photographic process and later calculated the size of the microplastic within a fixed pixel by converting between pixel and scale. The chemical composition of the samples was analyzed using a Fourier transform infrared microspectrometer (Nicolet iN 10 MX, Thermo Fisher, Waltham, MA, USA) transmission mode-MCT detector.

2.4. Contamination Mitigation

To prevent the samples from being contaminated by plastic fibers in the environment, cotton coats were worn during the analyses, and the surfaces of the operation platforms were wiped. All the glassware was thoroughly washed and covered with aluminum foil, and all prepared solutions were filtered through membranes before being used. The Petri dishes, filter membranes, and tweezers were inspected under a microscope to ensure that there was no microplastic pollution. When the laboratory digestion was conducted, an experimental blank was evaluated at the same time. The value of the blank was deducted when the results were calculated.

2.5. Statistical Analysis

The distribution of microplastics was mapped using the software Ocean Data View (5.1.5) and statistically analyzed and tabulated using Microsoft Office Excel 2016. The significant difference were measured via nonparametric tests (Mann–Whitney U-test). Moreover, we provide concentration methods for both the individual and mass benchmarks. The average density of microplastics used for the conversion in the paper was derived from Zhao [31].

3. Results

3.1. Seasonal Abundance Distribution of Microplastics

In this study, four surveys were carried out in the area off the Yangtze River estuary, and a total of 1825 plastic samples (with a size larger than 5 mm) and 1455 microplastic samples (with a size smaller than 5 mm) were analyzed, and microplastics accounted for 80% of the total. The average abundance of microplastics in the surface seawater off the Yangtze River estuary was (0.17 ± 0.14) items/m^3 (0.00561 ± 0.00462) mg/m^3. The average concentrations of microplastics in the four seasons of spring, summer, autumn, and winter were (0.05 ± 0.08) items/m^3 and (0.00165 ± 0.00264) mg/m^3, (0.25 ± 0.14) items/m^3 and (0.00825 ± 0.00462) mg/m^3, and (0.15 ± 0.13) items/m^3 (0.00495 ± 0.00429) mg/m^3 and (0.21 ± 0.11) items/m^3 (0.00693 ± 0.00363) mg/m^3, respectively. The concentrations of microplastics in the summer were higher than that in other seasons, and the abundance of microplastics in different seasons was in the same order of magnitude (Figure 2). The spatial distribution of microplastics shows that the nearshore abundance is higher than the farshore abundance in the sea area off the Yangtze River estuary.

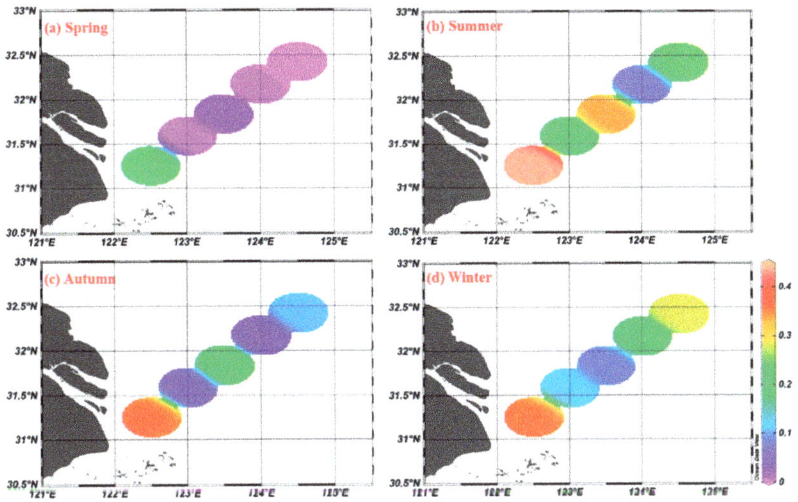

Figure 2. Distribution of microplastics in surface seawater of different season (items/m^3).

3.2. Particle Size, Shape, Color, and Composition of Microplastics

The physicochemical characteristics of microplastics in the surface seawater off the Yangtze River estuary are shown in Figures 3 and 4. The results show that microplastics with particle size in the range of (1–2) mm are the most numerous, accounting for 36.1% of all microplastics; the number of microplastics tends to increase with the decrease in particle size; and plastics tend to be miniaturized in the ocean, which is consistent with other studies [32–34]. The main shapes of microplastics are fiber, flake, and line, which accounted for 39.5%, 28.4%, and 20.8%, respectively, and the main compositions are polypropylene, polyethylene terephthalate, and polyethylene, accounting for 41.0%, 25.1%, and 24.9%, respectively. Yellow and green microplastics were the most abundant, accounting for 21.9% and 19.6%, respectively, followed by black (16.5%), transparent (15.7%), white (10.7%), and red (10.4%).

3.3. Characterization of Seasonal Patterns of Microplastics

In spring, microplastics with a particle size of less than 1 mm were the most abundant, accounting for 41.0% of the total, followed by (1–2) mm, accounting for 39.3%. The main shapes of microplastics were flake and line, accounting for 72.8% and 16.2%, respectively; the main compositions were polypropylene and polyethylene, accounting for 71.0% and 20.0%, respectively; and yellow and green microplastics were the most numerous, accounting for 63.8% and 17.2%.

In summer, microplastics with particle size (1–2) mm were the most numerous, accounting for 34.7%. The main shapes of microplastics were fibrous and linear, accounting for 68.8% and 17.1%, respectively; the main ingredient was polyethylene terephthalate, accounting for 69.2%; and the largest number of green and black microplastics were found, accounting for 37.0% and 26.0%, respectively.

In autumn, microplastics with particle size ranges of less than 1 mm and (1–2) mm accounted for the largest number of microplastics, 31.9% and 36.8%, respectively; the main shapes of microplastics were fiber and line, 52.4% and 20.7%, respectively; the main compositions of microplastics were polyethylene and polypropylene, 41.0% and 29.5%, respectively; and the largest number of transparent, white, yellow, and black microplastics were found, 32.6%, 32.6%, 16.0%, and 16.6%, respectively (Figure 5).

In winter, microplastics with particle size ranges of less than 1 mm and (1–2) mm accounted for the largest number of microplastics, with 22.1% and 34.2%, respectively; the main shapes of microplastics were fiber, line, and flake, with 33.7%, 26.1%, and 26.1%, respectively; and the main compositions were polyethylene, polyethylene terephthalate, and polypropylene, with 48.2%, 29.5%, and 17.6%, respectively; red, black, green, and blue microplastics were the most numerous, accounting for 25.9%, 21.7%, 17.1%, and 13.7%.

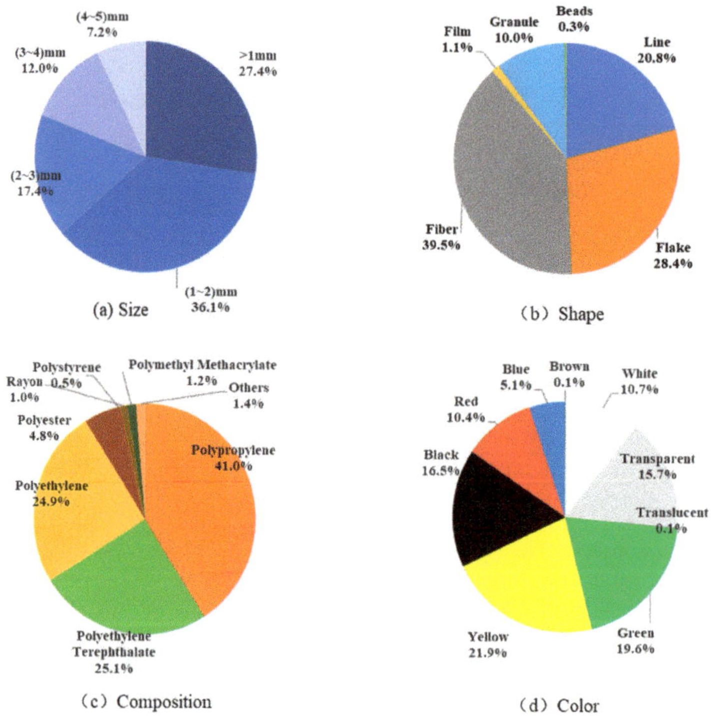

Figure 3. Physical and chemical characteristics of microplastics in the surface seawater off the Yangtze River estuary.

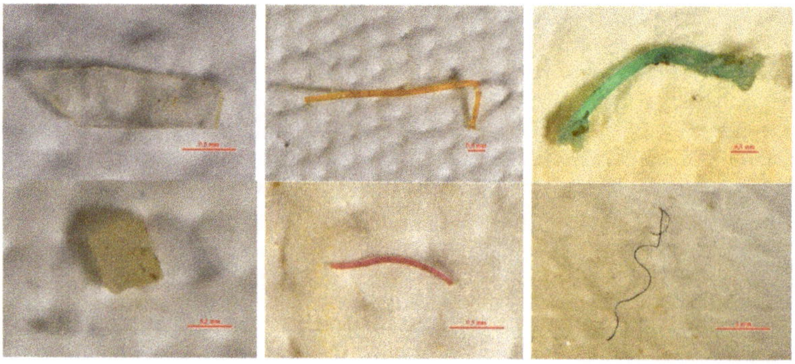

Figure 4. The types of microplastics in the surface seawater.

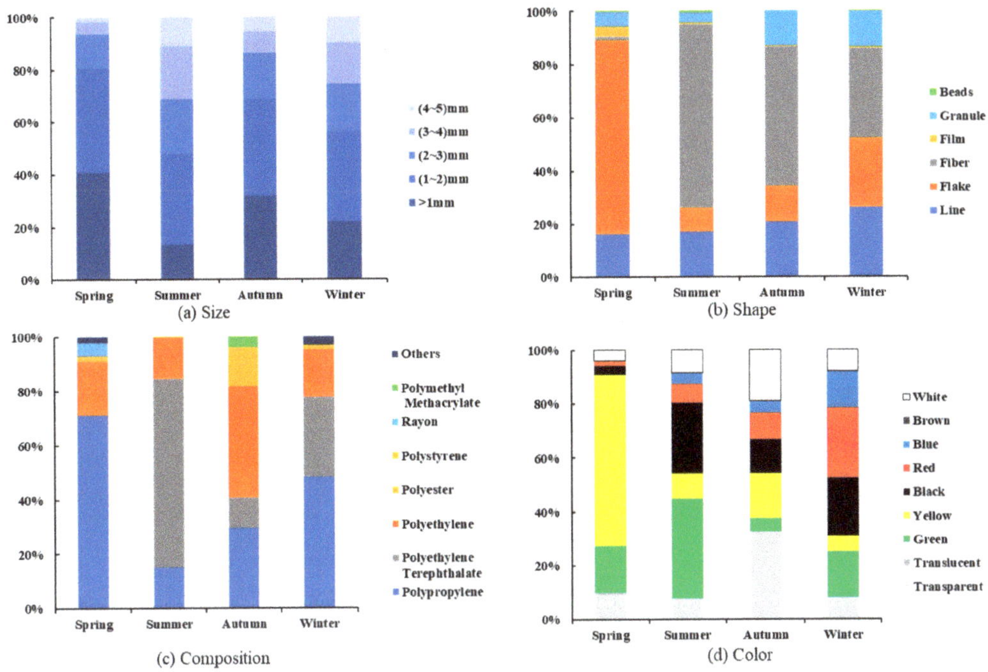

Figure 5. Physical and chemical characteristics of microplastics in different seasons.

4. Discussion

4.1. Levels of Microplastic Pollution in the Sea off the Yangtze River Estuary

Although microplastic monitoring has long been carried out in both domestic and international countries, there is no uniform standard yet. Different scholars' microplastic research methods are different, and the research differences are mainly the size of microplastic particles. For example, the particle sizes studied are (0.5~5) mm [26,27] and (0.05~3) mm [35], while the particle sizes of (0.33~5) mm [32,36–41] are the most studied. In order to assess the pollution level of microplastics in the sea area off the Yangtze River estuary, this paper compares with the results of foreign published studies on microplastics using nets of similar pore size.

The average abundance of microplastics in the waters off the Yangtze River estuary (0.17 items/m^3) (0.00561 mg/m^3) is in the same order of magnitude as that in the northwestern Mediterranean Sea [37], the Arctic Sea [38], and the Chukchi Sea [39] and is lower than that in the North Pacific Ocean [40], the California Sea [41,42], and the waters of the southeastern coast of South Korea [43], which suggests that the microplastics in the waters off the Yangtze River are at a moderately low level when compared to that of existing studies in foreign countries. Compared with the results of our coastal survey, it is in the same order of magnitude as the East China Sea Coastal [26], Bohai Sea [32], Jiangsu coastal area [44], Hangzhou Bay [45], Beibu Gulf [23], and Jinzhou Bay [46] and is higher than that of the South China Sea [47] and lower than that of Xiangshan Bay [48]. The level of microplastic pollution in the sea area off the Yangtze River estuary was at a moderately low level compared with domestic and international regions (Table 1).

Table 1. Comparison of microplastics between this study and other sampled areas.

Study Area	Net Mesh (μm)	Density (Items·m^{-3})	Study Period	Author
North Pacific Ocean circulation area	333	2.23	1999.8	Moore et al., 2001 [40]
Southern California coast	333	7.25	2000.10, 2001.1	Moore et al., 2002 [41]
Santa Monica Bay	333	3.92	2001.3	Lattin et al., 2004 [42]
Northwestern Mediterranean	333	0.116	2010.6–2010.9	Collignon et al., 2012 [37]
Arctic Sea	333	0.34 ± 0.31	2014.6	Lusher et al., 2015 [38]
Southeast coast of Korea	330	1.92–5.51	2012.5–2012.6, 2013.6–2013.7	Kang et al., 2015 [43]
Chukchi Sea	333	0.23 ± 0.07	2017.10	Mu et al., 2019 [39]
East China Sea coast	333	0.167 ± 0.138	2013.7–2013.8	Zhao et al., 2014 [26]
Bohai Sea	330	0.33 ± 0.34	2016.8	Zhang et al., 2017 [49]
South China Sea	333	0.045 ± 0.093	2017.4	Cai et al., 2018 [47]
Xiangshan Bay	333	8.91 ± 4.70	2017.10	Chen et al., 2018 [48]
Hangzhou Bay	330	0.14 ± 0.12	2019.10	Wang et al., 2020 [45]
Jinzhou Bay	330	0.65 ± 0.58	2016.10	Zhang et al., 2021 [46]
Beibu Gulf	300	0.1 ± 4.6	2018.10	Zhang et al., 2021 [23]
Jiangsu coastal area	200	3.1–3.5	2021.11	Xu et al., 2021 [44]
Off the Yangtze River estuary	330	0.17 ± 0.14	/	This study

4.2. Spatial and Temporal Distribution Factors of Microplastics off the Yangtze River Estuary

Microplastics were detected in different seasons at all stations except for station CJK5 in spring, where no microplastics were detected. Microplastics are commonly found in the sea area off the Yangtze River estuary, but their distribution is heterogeneous. The distribution of microplastics in seawater is affected by a variety of reasons, such as economic level, population density, circulation, wind, estuaries, harbors, and coastal sewage treatment plants in the countries along the ocean [50]. It was found that the abundance of microplastics at different stations in the sea area off the Yangtze River estuary varied greatly, and the abundance at nearshore stations was significantly higher than that in the farshore area. The overall spatial distribution of microplastics off the Yangtze River estuary showed a trend of higher abundance closer to the shore and lower abundance farther from the shore. In particular, the four-season abundance at station CJK1 was significantly higher than at the other stations, with an average abundance of 0.34 items/m^3 (0.01122 mg/m^3), while other stations yielded less than 0.2 items/m^3 (0.0066 mg/m^3). It is not difficult to see that the higher density of microplastics in the nearshore is mainly affected by the input from land sources, and the runoff of the Yangtze River has a decisive influence on the distribution of microplastics in the sea area outside the Yangtze River estuary. Preliminary analysis suggests that the distribution of microplastics in the sea area outside the Yangtze River estuary is mainly from riverine input in the nearshore, while riverine input is the main source of microplastics in the farshore, and human activities such as fishery fishing and shipping activities have aggravated the generation, aggregation, and diffusion of microplastics.

The reasons for the high level of microplastics in the nearshore and low level in the farshore of the Yangtze River estuary are manifold: (1) The study area is close to Shanghai and Jiangsu; the regional economy is developed; regional commerce, tourism, aquaculture, shipping, ports, and other activities are prosperous; the population is large; the amount of plastic waste generated by human activities is huge; and a large number of land-based sources of plastic garbage for the sea microplastics provide a rich material base conditions. (2) The spatial distribution of microplastics is influenced by hydrology and river input. The Yangtze River is the richest river in China, accounting for about 36% of the total river runoff, and the runoff of the Yangtze River provides important power conditions for the diffusion of microplastics into the sea. Microplastics from land-based sources are more likely to migrate with the Yangtze River runoff, which becomes one of the main reasons for the higher microplastics in the nearshore sea. Plastics and microplastic particles enter the sea with river water, the seawater has a dilution effect on the concentration of microplastics, and the dilution effect is intensified with the increase in the transportation distance of the Yangtze River freshwater, which results in the farshore site being lower than the nearshore site.

The study area is located at the mouth of the Yangtze River, which is strongly influenced by the river, and there are obvious seasonal variations in microplastic abundance in surface waters (Figure 6). The abundance of microplastics in the surface waters off the Yangtze River estuary was significantly higher in summer than in spring, followed by winter and autumn. The high abundance of microplastics in summer is mainly due to the influence of the East Asian monsoon and the abundant precipitation, which leads to maximum runoff of the Yangtze River into the sea throughout the year.

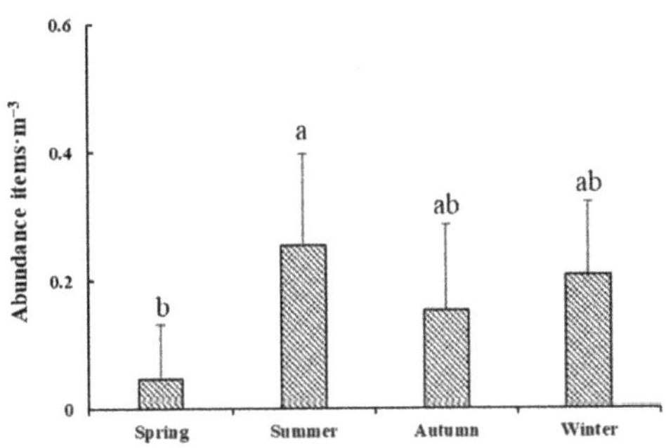

Figure 6. Comparison of microplastic abundance in different seasons. The letter indicated significant difference in Mann–Whitney U-test ($p < 0.05$).

Higher microplastic abundance was found in CJK5 and CJK4 at farshore stations in both fall and winter, which was significantly higher in winter than in autumn for several reasons. (1) Increased human activities: Frequent fishing activities occur after end of the fishing season in summer, and human activities influence the abundance and distribution of microplastics at sea to some extent, making higher concentrations in farshore areas possible. (2) Influence of the monsoon: Seasonal changes result in the summer prevailing southeasterly winds change to northerly winds, and the wind promotes the entry of plastic waste into the sea but also increases the content of microplastics in the ocean; on the other hand, the wind promotes the mixing of plastic debris in the ocean's upper waters and vertical redistribution, which may lead to increased microplastic content. (3) The impact of runoff is weakening: The runoff volume of the Yangtze River into the sea gradually decreases after the autumn, and the fresh water flushing from the Yangtze River shifts from the northeast to the southeast, and the region is basically unaffected by the fresh water flushing from the Yangtze River in winter. (4) Cumulative effects: With the cumulative effects of human activities and wind, microplastic abundance is higher in the farshore region in winter than in autumn.

Lower levels of microplastics were found in the farshore region in spring, especially in CJK5, where no microplastics were detected. Due to seasonal changes, the runoff from the Yangtze River gradually increases, the direction of freshwater flushing shifts from the southeast to the northeast, and the prevailing wind shifts from northerly to southeasterly, so microplastics are less likely to accumulate in this area, and thus, the abundance is lower relative to fall and winter.

4.3. Analysis of Sources of Microplastic Pollution

The source of marine microplastics is a difficult point in the study of the microplastics under research. Shape and material are important ways to determine the source, and this study attempts to analyze the possible sources of surface seawater microplastics in the waters off the Yangtze River estuary based on the shape and composition of the plastics (see

Figure 7). In the surface seawater microplastic samples, fibrous polyethylene terephthalate (24.2%), flaky polypropylene (21.7%), linear polypropylene (9.7%), and polyethylene (9.1%) were the main compositions of microplastics.

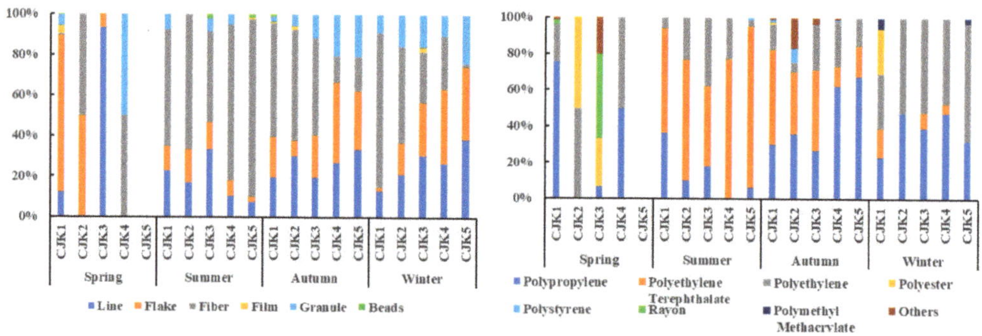

Figure 7. Types and components distribution of microplastics at sites.

Polypropylene is a thermoplastic polymer obtained by polymerization of propylene monomers, with a high melting point, high strength, high heat resistance, high abrasion resistance, and low creep but also with good tensile and yield strength, rigidity, stress resistance, and electrical insulation and other excellent properties. Polypropylene is used in granulated polypropylene, DVD packaging materials, polypropylene materials, random polypropylene-modified asphalt waterproofing materials, washing machine materials, cast polypropylene film materials, automotive plastics materials, and woven products made with polypropylene. In daily life, it is widely used in the packaging of clothing, textiles, bread, and other goods and also used in cable production [51]. From the shape analysis of flake polypropylene, mainly from the broken plastic packaging bags, the color is mainly yellow; the greatest number was found in spring, and some of the samples had become discolored or even cracked due to prolonged immersion in seawater. Linear polypropylene was mainly derived from the breaking of harbor shipping and fishing cables and nets, and the colors were mainly yellow, white, and red.

Polyethylene is the most widely used of the polymer materials, mainly used to manufacture film, packaging materials, containers, pipes, monofilaments, wires, cables, daily necessities, etc., and can be used as a high-frequency insulating materials for television, radar, etc. Analyzed in terms of shape, linear polyethylene originates from the breakage of cables and nets in port shipping and fishing. The color is predominantly green. All of these plastics have various degrees of deterioration.

Polyethylene terephthalate (PET) is a thermoplastic obtained by polycondensation of terephthalic acid (TPA) and ethylene glycol (EG) or prepared by ester exchange of dimethyl terephthalate (DMT) and EG. PET has the advantages of non-toxicity, tastelessness, light weight, high transparency, and better mechanical properties, so it is widely used in the fields of food packaging, fibers, and electrical insulating materials and other fields [52]. PET fiber is the number one chemical fiber species, and its production accounts for more than 80% of the total production of chemical fibers [53]. Analysis of fibrous PET from the main samples showed that it is produced in the processing, use, and cleaning process of textiles and has a color variety, mainly black, red, transparent, and blue. Some studies have shown that the number of microplastics released per 3 g of fabric washing can be up to 1300–1500 roots [54].

The survey also found a small amount of polyester and rayon fibers, which came from the same source as PET fibers, and a small amount of polystyrene foam, which was analyzed to have originated from foam rafts used in nearshore aquaculture and fishing activities.

In spring, microplastics were mainly flake polypropylene (65.5%) and line polyethylene (8.6%). In summer, microplastics were mainly fiber polyethylene terephthalate (68.4%), line polypropylene (8.0%), and polyethylene (8.7%). In autumn, microplastics were predominantly fiber polyethylene (18.1%), polyester (14.3%), polyethylene terephthalate (9.3%), polypropylene (9.0%), line polyethylene (13.0%), and polypropylene (6.2%). In winter, it was mainly fiber polyethylene terephthalate (29.0%), line polypropylene 19.4%), and flake polypropylene (17.0%). Preliminary analysis deduces that plastic packaging materials enter the sea more in spring, followed by discarded cables from fisheries. Domestic water discharges such as sewage treatment plants and nearshore fishing activities are the main sources of microplastics in summer and autumn, and sources of microplastics are diverse in winter.

4.4. Ecological Risk Assessment of Microplastics

The ecological risk index method [55] not only takes into account the impacts of various pollutants on the environment in a particular depositional environment but also adequately reflects the combined effects of multiple pollutants in the environment to quantitatively classify the potential ecological risk level; thus, it is one of the important methods of assessing the potential ecological risk of sediments. Peng [56] improved on the basis of the traditional potential ecological risk index method to study the risk of microplastic contamination in the surface water of pump mining in the Yangtze River estuary. In this study, we attempted to use its method to assess and study the potential risk of microplastics in surface waters off the Yangtze River estuary.

The formula is as follows:

$$C_f^i = \frac{C^i}{C_r^i} \quad T_r^i = \frac{P_i}{C^i} \times S_i \quad E_r^i = T_r^i \times C_f^i \quad RI = \sum_{i=1}^{n} E_r^i$$

C_f^i is the pollution index of microplastics, C^i is the measured concentration, C_r^i is the standard reference value—here, we refer to the safe concentration of microplastics in surface waters of 6650 particles/m^3 estimated by Everaert et al. [57]. We refer to Lithner et al. [58] to define the hazard index of microplastic polymers S_i; P_i is the concentration of specific microplastic polymer i concentration; T_r^i is the ecotoxicity response factor (the sum of the product of the percentage of the total amount of plastic polymer i and the hazard index of that polymer); E_r^i is the potential ecological risk index; RI is the ecological risk index of the microplastic polymers; n is the number of microplastic polymer species contained in the sample. We have listed the risk rank of polymers based on monomer toxicity in the Supplementary Material (Supplementary Material Table S1).

In terms of the pollution level of single-factor pollutants, the values of the abundance range of microplastics were non-detected to 0.44 items/m^3 (0.0132 mg/m^3), and the mean value was 0.17 items/m^3 (0.00561 mg/m^3), which was much lower than the safe concentration predicted by Everaert et al., indicating that the current status of microplastic pollution in the region is relatively light. In terms of potential ecological risk, the degree of microplastic contamination in the surface water network samples off the Yangtze River estuary was generally low, and the variation of the potential risk index (E_r^i) of each seasonal station ranged from 0 to 3.12×10^{-4}, which indicated that the microplastic contamination status in the surface seawater off the Yangtze River estuary was relatively light (Table 2).

Table 2. Risk-level criteria for microplastic pollution.

Potential Ecological Risk Factor E_r^i	<10	10–100	100–1000	>1000
Ecological Risk Scale for Microplastic Pollution	I	II	III	IV

It has been found that microplastics are easier to distribute in the natural environment due to their small size compared to large plastics. At the same time, due to the large specific surface area of microplastics in the water column, they can easily combine with other substances to form larger combinations and accumulate in sediments. Therefore, the abundance of microplastics in sediments is much higher than in water bodies, and the environmental conditions of sediments are more complex, resulting in greater potential hazards of microplastics in sediments.

Zhang et al.'s [59] study of sediments in the Yangtze River estuary showed that sediments were the main sink for microplastics, which tended to be distributed vertically in the upper sediments and suspended phases after suspended sedimentation [60].The abundance of microplastics in the upper sediment was significantly higher than that in the lower sediment. Meanwhile, the distribution of microplastics with different densities in the vertical space was different; low-density microplastics float in the water usually easily float in the water, and those with higher densities more easily sink. Immobilized microplastics in sediments may be reactivated by disturbances at the water–sediment interface [61], allowing them to migrate upward into the overlying water. At the same time, biological effects can also cause vertical spatial changes in microplastics. After being accidentally ingested by an organism, the already-deposited microplastics may undergo changes through the digestive system of the organism, leading to changes in their spatial distribution. In the horizontal direction, the distribution of microplastic abundance in sediments is roughly the same as that in the water column, with high volume areas occurring mainly in shallow, nearshore waters. The reasons affecting the horizontal distribution of microplastics can be summarized as follows: First, due to the scouring effect of the river, microplastics in the center of the river are washed to the riverbank and deposited. Secondly, due to the lesser flow of shallow water close to the shore, microplastics are easily deposited. Third, low-density microplastics can easily move with the monsoon and be deposited on lakeshore sediments near estuaries. Fourth, human activities are intensive in coastal areas, and the spatial variability of microplastics is mainly influenced by the extent, path, and location of human activities [62]; thus, the abundance of microplastics is relatively great in shallow waters near the coast. In ecosystems, biological uptake is a key aspect of the microplastic transport process, and microplastics in sediments are taken up by a large number of aquatic organisms, including fish [63]. Microplastics have been reported to be widely detected in various organisms, such as fish [64,65]. Feng et al. [66] demonstrated that MPs can accumulate in the gills, intestines, and skin of fish; Su et al. [67] investigated microplastics concentrations in fish from the Yangtze River estuary and found that MPs accumulated in the gills and intestines but not in the liver or muscles of fish; similarly, MPs were found in the gills, stomach, and intestines of fish from the mangrove wetlands of Zhanjiang [68]; Song et al. [69] also found that the detection rate of microplastics in wild fish from Haizhou Bay, Lvshi, and the Yangtze River estuary reached 98%, with concentrations of 0.28 ± 0.23 items/g, and MP abundance was highest in the skin (1.40 ± 1.38 items/individual), followed by gills (1.23 ± 1.07 items/individual), intestine (0.90 ± 0.95 items/individual), and liver (0.72 ± 0.91 items/individual). The intake of microplastics by fish not only limits their growth but may also jeopardize human health through the food chain [70], causing significant damage.

5. Conclusions

(1) The abundance of microplastics in the surface seawater off the Yangtze River estuary is (0.17 ± 0.14) items/m^3 (0.00561 ± 0.00462) mg/m^3, which is at a moderately low level compared with other regions both domestic and abroad;

(2) In surface seawater, the number of microplastics with a particle size of (1~2) mm is the largest; their shapes are mainly fibrous, flaky, and linear; their compositions are mainly polypropylene, polyethylene terephthalate, and polyethylene; and their colors are varied, mainly yellow and green;

(3) Influenced by land-based inputs, hydrodynamics, and human activities, the spatial distribution of microplastics in the surface seawater off the Yangtze River estuary is uneven, with a high level near the shore and a low level far away from the shore; there are obvious seasonal variations, with a higher level of microplastic pollution in

the summer; and the level of distribution is affected by the runoff significantly, with a higher level of pollution near the shore;

(4) Preliminary analysis suggests that marine shipping, fishing, and land-based sewage activities are important sources of microplastics in the waters off the Yangtze River estuary;

(5) The ecological risk index method can fully reflect the combined effects of multiple pollutants in the environment. The potential ecological risk of microplastics in the surface seawater off the Yangtze River estuary is small.

Supplementary Materials: The following supporting information can be downloaded at: https://www.mdpi.com/article/10.3390/toxics11110889/s1, Table S1. Risk rank of polymers based on monomer toxicity.

Author Contributions: Conceptualization, X.J. and H.L.; methodology, X.J.; formal analysis, X.J. and S.Y.; investigation, H.H.; data curation, X.J. and H.H.; writing—original draft preparation, X.J.; writing—review and editing, X.J. and H.L.; supervision, Y.H.; project administration, H.L.; funding acquisition, H.L. All authors have read and agreed to the published version of the manuscript.

Funding: This research was funded by the National Key Research and Development Program (2016YFC1402201) and the Science and Technology Innovation Program of Shanghai Municipal Science and Technology Commission (21DZ1201700, 22DZ1202702).

Institutional Review Board Statement: Not applicable.

Informed Consent Statement: Not applicable.

Data Availability Statement: Not applicable.

Acknowledgments: This work was supported by the MNR Key Laboratory of Marine Ecological Monitoring and Restoration Technology.

Conflicts of Interest: The authors declare no conflict of interest.

References

1. Thompson, R.C.; Olsen, Y.; Mitchell, R.P.; Davis, A.; Rowland, S.J.; John, A.W.G.; McGonigle, D.; Russell, A.E. Lost at Sea: Where Is All the Plastic? *Science* **2004**, *304*, 838. [CrossRef]
2. Sruthy, S.; Ramasamy, E.V. Microplastic pollution in Vembanad Lake, Kerala, India: The first report of microplastics in lake and estuarine sediments in India. *Environ. Pollut.* **2017**, *222*, 315–322. [CrossRef]
3. Wright, S.L.; Thompson, R.C.; Galloway, T.S. The physical impacts of microplastics on marine organisms: A review. *Environ. Pollut.* **2013**, *178*, 483–492. [CrossRef] [PubMed]
4. Watts, A.J.; Urbina, M.A.; Goodhead, R.; Moger, J.; Lewis, C.; Galloway, T.S. Effect of Microplastic on the Gills of the Shore Crab Carcinus maenas. *Environ. Sci. Technol.* **2016**, *50*, 5364–5369. [CrossRef] [PubMed]
5. Wang, B.; Yang, R.; Fang, Q. Marine plastic management policy agenda-setting in China (1985–2021): The Multi-stage Streams Framework. *Ocean Coast. Manag.* **2023**, *243*, 106761. [CrossRef]
6. Cózar, A.; Echevarría, F.; González-Gordillo, J.I.; Irigoien, X.; Úbeda, B.; Hernández-León, S.; Palma, Á.T.; Navarro, S.; García-de-Lomas, J.; Ruiz, A.; et al. Plastic debris in the open ocean. *Proc. Natl. Acad. Sci. USA* **2014**, *111*, 10239–10244. [CrossRef] [PubMed]
7. Auta, H.S.; Emenike, C.U.; Fauziah, S.H. Distribution and importance of microplastics in the marine environment: A review of the sources, fate, effects, and potential solutions. *Environ. Int.* **2017**, *102*, 165–176. [CrossRef] [PubMed]
8. Eriksen, M.; Mason, S.; Wilson, S.; Box, C.; Zellers, A.; Edwards, W.; Farley, H.; Amato, S. Microplastic pollution in the surface waters of the Laurentian Great Lakes. *Mar. Pollut. Bull.* **2013**, *77*, 177–182. [CrossRef]
9. Law, K.L.; Morét-Ferguson, S.E.; Goodwin, D.S.; Zettler, E.R.; DeForce, E.; Kukulka, T.; Proskurowski, G. Distribution of Surface Plastic Debris in the Eastern Pacific Ocean from an 11-Year Data Set. *Environ. Sci. Technol.* **2014**, *48*, 4732–4738. [CrossRef]
10. Law, K.L.; Morét-Ferguson, S.; Maximenko, N.A.; Proskurowski, G.; Peacock, E.E.; Hafner, J.; Reddy, C.M. Plastic accumulation in the North Atlantic subtropical gyre. *Science* **2010**, *329*, 1185–1188. [CrossRef]
11. Dubaish, F.; Liebezeit, G. Suspended Microplastics and Black Carbon Particles in the Jade System, Southern North Sea. *Water Air Soil Pollut.* **2013**, *224*, 1352. [CrossRef]
12. Sadri, S.S.; Thompson, R.C. On the quantity and composition of floating plastic debris entering and leaving the Tamar Estuary, Southwest England. *Mar. Pollut. Bull.* **2014**, *81*, 55–60. [CrossRef] [PubMed]
13. Van Cauwenberghe, L.; Vanreusel, A.; Mees, J.; Janssen, C.R. Microplastic pollution in deep-sea sediments. *Environ. Pollut.* **2013**, *182*, 495–499. [CrossRef]

14. Zhang, K.; Su, J.; Xiong, X.; Wu, X.; Wu, C.; Liu, J. Microplastic pollution of lakeshore sediments from remote lakes in Tibet plateau, China. *Environ. Pollut.* **2016**, *219*, 450–455. [CrossRef]
15. Bergmann, M.; Sandhop, N.; Schewe, I.; D'Hert, D. Observations of floating anthropogenic litter in the Barents Sea and Fram Strait, Arctic. *Polar Biol.* **2016**, *39*, 553–560. [CrossRef]
16. Cincinelli, A.; Scopetani, C.; Chelazzi, D.; Lombardini, E.; Martellini, T.; Katsoyiannis, A.; Fossi, M.C.; Corsolini, S. Microplastic in the surface waters of the Ross Sea (Antarctica): Occurrence, distribution and characterization by FTIR. *Chemosphere* **2017**, *175*, 391–400. [CrossRef]
17. Carpenter, E.J.; Smith, K.L., Jr. Plastics on the Sargasso sea surface. *Science* **1972**, *175*, 1240–1241. [CrossRef]
18. Masó, M.; Garcés, E.; Pagès, F.; Camp, J. Drifting plastic debris as a potential vector for dispersing Harmful Algal Bloom (HAB) species. *Sci. Mar.* **2003**, *67*, 107–111. [CrossRef]
19. Votier, S.C.; Archibald, K.; Morgan, G.; Morgan, L. The use of plastic debris as nesting material by a colonial seabird and associated entanglement mortality. *Mar. Pollut. Bull.* **2011**, *62*, 168–172. [CrossRef]
20. Casale, P.; Affronte, M.; Insacco, G.; Freggi, D.; Vallini, C.; d'Astore, P.; Basso, R.; Paolillo, G.; Abbate, G.; Argano, R. Sea turtle stranding reveal high anthropogenic mortality in Italian waters. *Aquat. Conserv. Mar. Freshw. Ecosyst.* **2010**, *20*, 611–620. [CrossRef]
21. Yang, D.; Shi, H.; Li, L.; Li, J.; Jabeen, K.; Kolandhasamy, P. Microplastic Pollution in Table Salts from China. *Environ. Sci. Technol.* **2015**, *49*, 13622–13627. [CrossRef] [PubMed]
22. Galloway, T.S.; Lewis, C.N. Marine microplastics spell big problems for future generations. *Proc. Natl. Acad. Sci. USA* **2016**, *113*, 2331–2333. [CrossRef] [PubMed]
23. Zhang, L.; Liu, J.; Xie, Y.; Zhong, S.; Yang, B.; Lu, D.; Zhong, Q. Distribution of microplastics in surface water and sediments of Qin river in Beibu Gulf, China. *Sci. Total Environ.* **2020**, *708*, 135176. [CrossRef]
24. Zhao, S.; Zhu, L.; Li, D. Characterization of small plastic debris on tourism beaches around the South China Sea. *Reg. Stud. Mar. Sci.* **2015**, *1*, 55–62. [CrossRef]
25. Li, J.; Qu, X.; Su, L.; Zhang, W.; Yang, D.; Kolandhasamy, P.; Li, D.; Shi, H. Microplastics in mussels along the coastal waters of China. *Environ. Pollut.* **2016**, *214*, 177–184. [CrossRef] [PubMed]
26. Zhao, S.; Zhu, L.; Wang, T.; Li, D. Suspended microplastics in the surface water of the Yangtze Estuary System, China: First observations on occurrence, distribution. *Mar. Pollut. Bull.* **2014**, *86*, 562–568. [CrossRef] [PubMed]
27. Liu, T.; Sun, X.-X.; Zhu, M.L.; Liang, J.H.; Zhao, Y. Distribution and composition of microplastics in the surface water of the East China Sea. *Oceanol. Et Limnol. Sin.* **2018**, *49*, 62–69. [CrossRef]
28. Zhou, P.; Huang, C.; Fang, H.; Cai, W.; Li, D.; Li, X.; Yu, H. The abundance, composition and sources of marine debris in coastal seawaters or beaches around the northern South China Sea (China). *Mar. Pollut. Bull.* **2011**, *62*, 1998–2007. [CrossRef]
29. Tsang, Y.Y.; Mak, C.W.; Liebich, C.; Lam, S.W.; Sze, E.T.P.; Chan, K.M. Microplastic pollution in the marine waters and sediments of Hong Kong. *Mar. Pollut. Bull.* **2017**, *115*, 20–28. [CrossRef]
30. Kunz, A.; Walther, B.A.; Löwemark, L.; Lee, Y.-C. Distribution and quantity of microplastic on sandy beaches along the northern coast of Taiwan. *Mar. Pollut. Bull.* **2016**, *111*, 126–135. [CrossRef]
31. Zhao, S.; Wang, T.; Zhu, L.; Xu, P.; Wang, X.; Gao, L.; Li, D. Analysis of suspended microplastics in the Changjiang Estuary: Implications for riverine plastic load to the ocean. *Water Res.* **2019**, *161*, 560–569. [CrossRef]
32. Shu, X.; Xu, L.; Yang, M.; Qin, Z.; Zhang, Q.; Zhang, L. Spatial distribution characteristics and migration of microplastics in surface water, groundwater and sediment in karst areas: The case of Yulong River in Guilin, Southwest China. *Sci. Total Environ.* **2023**, *868*, 161578. [CrossRef] [PubMed]
33. Barnes, D.K.; Galgani, F.; Thompson, R.C.; Barlaz, M. Accumulation and fragmentation of plastic debris in global environments. *Philos. Trans. R. Soc. B Biol. Sci.* **2009**, *364*, 1985–1998. [CrossRef] [PubMed]
34. Song, Y.K.; Hong, S.H.; Jang, M.; Kang, J.-H.; Kwon, O.Y.; Han, G.M.; Shim, W.J. Large Accumulation of Micro-sized Synthetic Polymer Particles in the Sea Surface Microlayer. *Environ. Sci. Technol.* **2014**, *48*, 9014–9021. [CrossRef] [PubMed]
35. Hidalgo-Ruz, V.; Gutow, L.; Thompson, R.C.; Thiel, M. Microplastics in the Marine Environment: A Review of the Methods Used for Identification and Quantification. *Environ. Sci. Technol.* **2012**, *46*, 3060–3075. [CrossRef]
36. Yamashita, R.; Tanimura, A. Floating plastic in the Kuroshio Current area, western North Pacific Ocean. *Mar. Pollut. Bull.* **2007**, *54*, 485–488. [CrossRef]
37. Collignon, A.; Hecq, J.-H.; Glagani, F.; Voisin, P.; Collard, F.; Goffart, A. Neustonic microplastic and zooplankton in the North Western Mediterranean Sea. *Mar. Pollut. Bull.* **2012**, *64*, 861–864. [CrossRef]
38. Lusher, A.L.; Tirelli, V.; O'Connor, I.; Officer, R. Microplastics in Arctic polar waters: The first reported values of particles in surface and sub-surface samples. *Sci. Rep.* **2015**, *5*, 14947. [CrossRef]
39. Mu, J.; Zhang, S.; Qu, L.; Jin, F.; Fang, C.; Ma, X.; Zhang, W.; Wang, J. Microplastics abundance and characteristics in surface waters from the Northwest Pacific, the Bering Sea, and the Chukchi Sea. *Mar. Pollut. Bull.* **2019**, *143*, 58–65. [CrossRef]
40. Moore, C.J.; Moore, S.L.; Leecaster, M.K.; Weisberg, S.B. A Comparison of Plastic and Plankton in the North Pacific Central Gyre. *Mar. Pollut. Bull.* **2001**, *42*, 1297–1300. [CrossRef]
41. Moore, C.J.; Moore, S.L.; Weisberg, S.B.; Lattin, G.L.; Zellers, A.F. A comparison of neustonic plastic and zooplankton abundance in southern California's coastal waters. *Mar. Pollut. Bull.* **2002**, *44*, 1035–1038. [CrossRef] [PubMed]

42. Lattin, G.L.; Moore, C.J.; Zellers, A.F.; Moore, S.L.; Weisberg, S.B. A comparison of neustonic plastic and zooplankton at different depths near the southern California shore. *Mar. Pollut. Bull.* **2004**, *49*, 291–294. [CrossRef] [PubMed]
43. Kang, J.-H.; Kwon, O.-Y.; Shim, W.J. Potential Threat of Microplastics to Zooplanktivores in the Surface Waters of the Southern Sea of Korea. *Arch. Environ. Contam. Toxicol.* **2015**, *69*, 340–351. [CrossRef]
44. Xu, N.; Jiang, L.; Zhang, Y.; Shen, Y.; Wang, Y.; Wang, S.; Yuan, Q. Microplastic pollution in the offshore sea, rivers and wastewater treatment plants in Jiangsu coastal area in China. *Mar. Environ. Res.* **2023**, *188*, 105992. [CrossRef]
45. Wang, T.; Hu, M.; Song, L.; Yu, J.; Liu, R.; Wang, S.; Wang, Z.; Sokolova, I.M.; Huang, W.; Wang, Y. Coastal zone use influences the spatial distribution of microplastics in Hangzhou Bay, China. *Environ. Pollut.* **2020**, *266*, 115137. [CrossRef] [PubMed]
46. Zhang, W.; Zhang, S.; Zhao, Q.; Qu, L.; Ma, D.; Wang, J. Spatio-temporal distribution of plastic and microplastic debris in the surface water of the Bohai Sea, China. *Mar. Pollut. Bull.* **2020**, *158*, 111343. [CrossRef] [PubMed]
47. Cai, M.; He, H.; Liu, M.; Li, S.; Tang, G.; Wang, W.; Huang, P.; Wei, G.; Lin, Y.; Chen, B.; et al. Lost but can't be neglected: Huge quantities of small microplastics hide in the South China Sea. *Sci. Total Environ.* **2018**, *633*, 1206–1216. [CrossRef]
48. Chen, M.; Jin, M.; Tao, P.; Wang, Z.; Xie, W.; Yu, X.; Wang, K. Assessment of microplastics derived from mariculture in Xiangshan Bay, China. *Environ. Pollut.* **2018**, *242*, 1146–1156. [CrossRef]
49. Zhang, W.; Zhang, S.; Wang, J.; Wang, Y.; Mu, J.; Wang, P.; Lin, X.; Ma, D. Microplastic pollution in the surface waters of the Bohai Sea, China. *Environ. Pollut.* **2017**, *231*, 541–548. [CrossRef]
50. Renchen, G.F.; Butler, C.B.; Matthews, T.R. Marine debris knows no boundaries: Characteristics of debris accumulation in marine protected areas of the Florida Keys. *Mar. Pollut. Bull.* **2021**, *173*, 112957. [CrossRef]
51. Su, L.; Xiong, X.; Zhang, Y.; Wu, C.; Xu, X.; Sun, C.; Shi, H. Global transportation of plastics and microplastics: A critical review of pathways and influences. *Sci. Total Environ.* **2022**, *831*, 154884. [CrossRef] [PubMed]
52. Zhou, K.; Liu, Q.; Feng, J.; Chang, T.; Liu, J. Comprehensive environmental performance of bottle-to-bottle recycling of PET bottles based on deposit-refund system in China. *Waste Manag.* **2023**, *172*, 90–100. [CrossRef] [PubMed]
53. Pinlova, B.; Hufenus, R.; Nowack, B. Systematic study of the presence of microplastic fibers during polyester yarn production. *J. Clean. Prod.* **2022**, *363*, 132247. [CrossRef]
54. Dalla Fontana, G.; Mossotti, R.; Montarsolo, A. Assessment of microplastics release from polyester fabrics: The impact of different washing conditions. *Environ. Pollut.* **2020**, *264*, 113960. [CrossRef] [PubMed]
55. Hakanson, L. An ecological risk index for aquatic pollution control. A sedimentological approach. *Water Res.* **1980**, *14*, 975–1001. [CrossRef]
56. Peng, G.; Xu, P.; Zhu, B.; Bai, M.; Li, D. Microplastics in freshwater river sediments in Shanghai, China: A case study of risk assessment in mega-cities. *Environ. Pollut.* **2018**, *234*, 448–456. [CrossRef]
57. Everaert, G.; Van Cauwenberghe, L.; De Rijcke, M.; Koelmans, A.A.; Mees, J.; Vandegehuchte, M.; Janssen, C.R. Risk assessment of microplastics in the ocean: Modelling approach and first conclusions. *Environ. Pollut.* **2018**, *242*, 1930–1938. [CrossRef]
58. Lithner, D.; Larsson, Å.; Dave, G. Environmental and health hazard ranking and assessment of plastic polymers based on chemical composition. *Sci. Total Environ.* **2011**, *409*, 3309–3324. [CrossRef]
59. Zhang, B.; Chen, L.; Chao, J.; Yang, X.; Wang, Q. Research Progress of Microplastics in Freshwater Sediments in China. *Environ. Sci. Pollut. Res.* **2020**, *27*, 31046–31060. [CrossRef]
60. Zhu, X.-T.; Yi, J.; Qiang, L.-Y.; Cheng, J.-P. Distribution and Settlement of Microplastics in the Surface Sediment of Yangtze Estuary. *Huan Jing Ke Xue* **2018**, *39*, 2067–2074. [CrossRef]
61. Ji, X.; Ma, Y.; Zeng, G.; Xu, X.; Mei, K.; Wang, Z.; Chen, Z.; Dahlgren, R.; Zhang, M.; Shang, X. Transport and fate of microplastics from riverine sediment dredge piles: Implications for disposal. *J. Hazard. Mater.* **2021**, *404*, 124132. [CrossRef] [PubMed]
62. Ioka, S.; Muraoka, H.; Matsuyama, K.; Tomita, K. In situ redox potential measurements as a monitoring technique for hot spring water quality. *Sustain. Water Resour. Manag.* **2016**, *2*, 353–358. [CrossRef]
63. Zhang, T.; Sun, Y.; Song, K.; Du, W.; Huang, W.; Gu, Z.; Feng, Z. Microplastics in different tissues of wild crabs at three important fishing grounds in China. *Chemosphere* **2021**, *271*, 129479. [CrossRef]
64. McNeish, R.E.; Kim, L.H.; Barrett, H.A.; Mason, S.A.; Kelly, J.J.; Hoellein, T.J. Microplastic in riverine fish is connected to species traits. *Sci. Rep.* **2018**, *8*, 11639. [CrossRef]
65. Li, B.; Liang, W.; Liu, Q.-X.; Fu, S.; Ma, C.; Chen, Q.; Su, L.; Craig, N.J.; Shi, H. Fish Ingest Microplastics Unintentionally. *Environ. Sci. Technol.* **2021**, *55*, 10471–10479. [CrossRef]
66. Feng, Z.; Zhang, T.; Li, Y.; He, X.; Wang, R.; Xu, J.; Gao, G. The accumulation of microplastics in fish from an important fish farm and mariculture area, Haizhou Bay, China. *Sci. Total Environ.* **2019**, *696*, 133948. [CrossRef] [PubMed]
67. Su, L.; Deng, H.; Li, B.; Chen, Q.; Pettigrove, V.; Wu, C.; Shi, H. The occurrence of microplastic in specific organs in commercially caught fishes from coast and estuary area of east China. *J. Hazard Mater.* **2019**, *365*, 716–724. [CrossRef]
68. Huang, J.-S.; Koongolla, J.B.; Li, H.-X.; Lin, L.; Pan, Y.-F.; Liu, S.; He, W.-H.; Maharana, D.; Xu, X.-R. Microplastic accumulation in fish from Zhanjiang mangrove wetland, South China. *Sci. Total Environ.* **2020**, *708*, 134839. [CrossRef]

69. Song, K.; Du, W.; Ma, X.; Chen, Y.; Sun, Y.; Zhang, T.; Huang, W.; Feng, Z. Accumulation of microplastics in fugu (*Takifugu bimaculatus*): A comparative study between fishing grounds and aquafarms. *Mar. Pollut. Bull.* **2022**, *185*, 114200. [CrossRef]
70. Wu, F.; Wang, T.; Li, X.; Zhao, R.; He, F. Microplastic contamination in the dominant crabs at the intertidal zone of Chongming Island, Yangtze Estuary. *Sci. Total Environ.* **2023**, *896*, 165258. [CrossRef]

toxics

Article

Lutein Modulates Oxidative Stress, Inflammatory and Apoptotic Biomarkers Related to Di-(2-Ethylhexyl) Phthalate (DEHP) Hepato-Nephrotoxicity in Male Rats: Role of Nuclear Factor Kappa B

Dina R. S. Gad El-Karim [1], Mohamed A. Lebda [2], Badriyah S. Alotaibi [3,*], Attalla F. El-kott [4,5], Heba I. Ghamry [6] and Mustafa Shukry [7,*]

1 Department of Pathology and Clinical Pathology, Faculty of Veterinary Medicine, Alexandria University, Alexandria 22758, Egypt
2 Department of Biochemistry, Faculty of Veterinary Medicine, Alexandria University, Alexandria 22758, Egypt
3 Department of Pharmaceutical Sciences, College of Pharmacy, Princess Nourah bint Abdulrahman University, P.O. Box 84428, Riyadh 11671, Saudi Arabia
4 Department of Biology, College of Science, King Khalid University, Abha 61421, Saudi Arabia
5 Department of Zoology, Faculty of Science, Damanhour University, Damanhour 22511, Egypt
6 Nutrition and Food Sciences, Department of Home Economics, Faculty of Home Economics, King Khalid University, P.O. Box 960, Abha 61421, Saudi Arabia
7 Physiology Department, Faculty of Veterinary Medicine, Kafrelsheikh University, Kafrelsheikh 33516, Egypt
* Correspondence: badriyah.salotaibi@gmail.com (B.S.A.); mostafa.ataa@vet.kfs.edu.eg (M.S.)

Abstract: Phthalates are widely distributed in our environment due to their usage in many industries, especially in plastic production, which has become an essential part of daily life. This investigation aimed to assess the potential remedial influence of lutein, a naturally occurring carotenoid, on phthalate-triggered damage to the liver and kidneys. When di-(2-ethylhexyl) phthalate (DEHP) was administered to male albino rats over sixty straight days at a dosage of 200 mg/kg body weight, it resulted in a significant increase in the serum activity of liver enzymes (AST, ALT, and GGT), alpha-fetoprotein, creatinine, and cystatin-C, as well as disruptions in the serum protein profile. In addition, intoxication with DEHP affected hepato-renal tissues' redox balance. It increased the content of some proinflammatory cytokines, nuclear factor kappa B (Nf-κB), and apoptotic marker (caspase-3); likewise, DEHP-induced toxicity and decreased the level of anti-apoptotic protein (Bcl-2) in these tissues. Lutein administration at a dose level of 40 mg/kg b.w efficiently facilitated the changes in serum biochemical constituents, hepato-renal oxidative disturbance, and inflammatory, apoptotic, and histopathological alterations induced by DEHP intoxication. In conclusion, it can be presumed that lutein is protective as a natural carotenoid against DEHP toxicity.

Keywords: phthalates; lutein; hepato-nephrotoxicity; rats; Nf-κB

1. Introduction

Phthalates (phthalic acid derivatives) are synthetic chemicals that are widely produced for usage in many industries. They are the main part of the plastic industry as plasticizers due to their ability to enhance plastic materials' durability, transparency, and flexibility. So, the phthalates group is one of the most abundant environmental contaminants [1]. In addition, phthalates could be used as solvents in various products, including paints and insect sprays, and as color and scent stabilizers in cosmetics [2]. Massive usage of phthalates leads to exposure of humans and animals to their toxicity via ingestion, dermal absorption, and inhalation [3]. Still, ingestion is the main route of exposure [4]. Di-(2-ethylhexyl) phthalate (DEHP) is the main phthalate derivative that is used in the production of polyvinyl chloride (PVC). Unfortunately, DEHP leaks easily from PVC due to its weak bond with plastic material, contributing to its health hazard effect [5]. DEHP and its metabolites are linked with a wide range of adverse effects on the kidney, liver, heart, lung, and reproductive tract [6]. The presence of phthalate derivative metabolites was proved in all tested human urine samples by Heudorf et al. [7]. Different mechanisms are implicated in the induction of phthalates toxicity, but oxidative stress is the most acceptable [8]. The metabolism of DEHP occurs in various tissues containing hydrolase, unspecific lipase, and esterase enzymes, producing mono-2-ethylhexyl) phthalate and 2-ethylhexanol [9]. These metabolized products disrupt mitochondrial function and release cytochrome c, leading to the generation of ROS and decreased antioxidant capacity [10]. DEHP induces liver injury by promoting the production of ROS and activating LPO, resulting in lipid peroxidation and damage to the cell structure and function of the liver [11]. Zhang et al. [12] found that excessive ROS production due to DEHP treatment affects MDA and SOD levels, leading to cell lipid accumulation and potential cell death. Excessive ROS reacts with fatty acids to form LPO products like MDA and HNE, causing cell membrane fluidity and permeability changes, leading to cell structure and function damage [13]. On the other hand, lutein is a natural carotenoid present abundantly in green and colored vegetables and fruits [14]. Lutein has many pharmacological and biological properties, including antioxidant [15], anti-inflammatory [16], hepato-protective [17], nephroprotective [18], cardio-protective [19], and anti-cancer [20] effects. In this consistency, this study aimed to evaluate the toxic impact of di-(2-ethylhexyl) phthalate on the liver and kidney and the presumptive ameliorative role of lutein as a natural carotenoid against this toxicity.

2. Material and Methods

2.1. Chemicals

Di (2-ethylhexyl) phthalate (DEHP, ±99% purity) was purchased from Sigma–Aldrich (St Louis, MO, USA). Lutein was obtained as capsules containing 20 mg of lutein ((Des Moines, IA, USA). Biochemical diagnostic kits were from Spinreact Kits, Spain; Cusabio®, Wuhan, China; Crystal Chem, Elk Grove Village, IL, USA; and Biodiagnostic Co, Giza, Egypt.

2.2. Experimental Animals and Design

Twenty-eight male albino rats (180–200 g) were purchased from Pharos University Animal House, Alexandria, Egypt, to accomplish this study. They were kept in separate metal cages at ambient humidity and temperature; a 12/12 h light/darkness cycle was applied. Basal experimental diet and water were offered ad libitum; they were kept without any treatment for ten days to acclimatize and to ensure they were free from any apparent health problems. After acclimatization, they were divided into four equal groups (7 rats/each): the control group received corn oil 1 mL/kg orally; the lutein group received lutein at a dose level of 40 mg/kg b.w; DEHP group intubated with DEHP at a dose level of 200 mg/kg b.w; DEHP/lutein group received both DEHP (200 mg/kg b.w) [21–24] and lutein (40 mg/kg b.w) [25].

All the treatments were administrated orally after dissolving in corn oil using gastric gavage. The treatment protocols were administrated daily for sixty consecutive days. The animals were sacrificed through cervical dislocation after deep anesthesia with ketamine/xylazine (7.5–10 mg/kg, 1 mg/kg i.p) 24 h. following the last administration of the treatment protocols.

2.3. Sampling and Biochemical Analysis

2.3.1. Serum Biochemical Analysis

Blood samples were obtained from the heart directly using a syringe after dissection of the sacrificed animals. The blood was drained in plain tubes, left to coagulate for 30 min at room temperature, and then centrifuged at 3000 rpm for 10 min to separate serum. Serum aliquots were kept at −20 °C for further detection of serum activities of hepatic transaminase enzymes (AST and ALT), gamma-glutamyl transferase enzyme (GGT), in addition to the serum concentration of creatinine, total protein, and albumin (Spinreact kits, Barcelona, Spain). Also, serum levels of alpha-fetoprotein (AFP) (Cusabio®, Wuhan, China) and cystatin-C (Crystal Chem, 955 Busse Rd, Elk Grove Village, IL 60007, USA) were detected.

2.3.2. Preparation of Tissue Homogenate

The left kidney of each animal, in addition to one lobe of the liver, was obtained, washed several times with phosphate buffer saline (PBS), and jabbed with PBS containing heparin to remove any blood clots. Samples were then dissected into small pieces using a scalpel, and PBS was added to obtain 10% tissue homogenate using a tissue homogenizer (Glas-Col®, Beijing, China). The obtained homogenates were centrifuged at 3000 rpm for 20 min and filtrated; the clear supernatant was kept at −80 °C for further evaluation of oxidant/antioxidant, inflammatory, and apoptotic biomarkers. The protein content of homogenate was detected using Bradford's reagent Sigma-Aldrich (St. Louis, MO, USA).

2.3.3. Evaluation of Oxidant/Antioxidant Biomarkers

The levels of malondialdehyde (MDA), reduced glutathione (GSH), and the activity of the catalase enzyme (CAT) in the liver and kidneys were measured utilizing kits commercially sourced from Biodiagnostic, Egypt.

2.3.4. Evaluation of Inflammatory and Apoptotic Biomarkers

Hepatic and renal levels of interleukin-1beta (IL-1β), tumor necrosis factor-alpha (TNF-α) (Abcam, Cambridge, MA, USA), Bcl-2, caspase-3 and nuclear factor kappa B (Cusabio®, Wuhan, China) were determined using previously listed highly specific ELISA-based kits.

2.3.5. Histopathological Examination

The right kidney of each animal and small pieces of liver were removed and washed several times with normal saline and kept in 10% formalin solution to perform a 5 µm thickness tissue section for staining with H&E, according to Bancroft and Stevens [26]

2.3.6. Histopathological Semi-Quantitative Scoring System

Five random fields (×100) were randomly selected from each rat in each group. The present pathological lesions were detected and scored as follows according to the severity of the presented lesions, which were achieved relying on the percentage of the affected area/entire section: - = absence of lesion, + (mild) = 5-25%, ++ (moderate) = 26-50%, and +++ (severe) = ≥50%.

2.3.7. Statistical Analysis

A one-way analysis of variance (ANOVA) test, executed through the Statistical Analysis System (SAS) software, was employed to investigate the influence of distinct treatment methods on the parameters assessed.

3. Results

3.1. Serum Findings of Liver and Kidney Functions

As present in Table 1, serum activities of hepatic enzymes AST (+80%), ALT (+120%), and GGT (+311%), in addition to the serum level of AFP (+120%), globulin (+37%) and renal biomarker; cystatin-C (+150%) were significantly incriminated in DEHP-intoxicated rats, while serum level of albumin (−26%) was decreased in the same group when compared to control rats. These changes were facilitated significantly in rats pretreated with lutein, improving the liver and renal biomarkers.

Table 1. Serum biochemical findings of liver and kidney functions in DEHP and lutein-treated groups.

	Control	Lutein	DEHP	DEHP/Lutein
AST (U/L)	115.00 [c] ± 7.12	114.71 [c] ± 7.48	209.29 [a] ± 7.85	166.86 [b] ± 5.45
ALT (U/L)	24.14 [c] ± 2.26	23.57 [c] ± 1.86	53.29 [a] ± 3.39	41.00 [b] ± 1.98
GGT (U/L)	18.14 [c] ± 2.18	17.64 [c] ± 1.96	74.14 [a] ± 4.67	47.57 [b] ± 4.37
AFP (ng/mL)	5.11 [c] ± 0.64	4.90 [c] ± 0.55	11.50 [a] ± 1.15	7.54 [b] ± 0.47
Cystatin-C (mg/L)	2.58 [c] ± 0.33	2.47 [c] ± 0.29	5.82 [a] ± 0.40	3.87 [b] ± 0.35
Total protein (g/dL)	6.36 ± 0.12	6.37 ± 0.13	6.16 ± 0.15	6.41 ± 0.16
Albumin (g/dL)	3.83 [a] ± 0.13	3.87 [a] ± 0.13	2.80 [b] ± 0.13	3.51 [a] ± 0.14
Globulins (g/dL)	2.51 [c] ± 0.09	2.50 [c] ± 0.05	3.44 [a] ± 0.11	2.94 [b] ± 0.06

Results are displayed as Mean ± Standard Error. [a,b,c] Noteworthy differences ($p < 0.05$) are observed among the means within the same rows of distinct litters. AST; aspartate aminotransferase, ALT; alanine aminotransferase, GGT; gamma-glutamyl transferase, AFP; alfa fetoprotein, DEHP; di-(2-ethylhexyl) phthalate.

3.2. Hepato-Renal Redox State

Intoxication with DEHP significantly increased the content of lipid peroxide (MDA) in both hepatic (+87%) and renal (+138%) tissues, depleted the concentration of GSH (−50%, −36%, respectively), and the activity of CAT enzyme within these tissues in comparison with a control group. However, co-treatment with lutein restored the oxidative balance of these tissues partially, as the level of MDA was decreased (~23% in both tissues). The activity of CAT was increased (~58% in the liver and ~38% in the kidneys), alongside significant elevation in the GSH content (~55% in the liver and ~29% in kidneys), as represented in Table 2.

Table 2. Hepato-renal oxidant/antioxidant biomarkers in DEHP and lutein-treated groups.

	Control	Lutein	DEHP	DEHP/Lutein
Hepatic oxidant/antioxidative indices				
MDA (nmol/mg protein)	117.00 [c] ± 8.31	119.00 [c] ± 9.06	219.86 [a] ± 10.90	171.57 [b] ± 12.25
GSH (nmol/mg protein)	18.86 [a] ± 1.79	18.21 [a] ± 0.98	9.43 [c] ± 0.62	14.24 [b] ± 1.04
CAT (U/mg protein)	26.71 [a] ± 1.82	27.86 [a] ± 1.75	12.07 [c] ± 1.21	19.00 [b] ± 1.50
Renal oxidant/antioxidative indices				
MDA (nmol/mg protein)	50.57 [c] ± 3.64	52.14 [c] ± 3.62	119.14 [a] ± 7.23	91.86 [b] ± 4.61
GSH (nmol/mg protein)	44.43 [a] ± 2.40	45.00 [a] ± 1.94	28.14 [c] ± 2.59	36.29 [b] ± 1.96
CAT (U/mg protein)	4.20 [a] ± 0.45	4.23 [a] ± 0.50	2.61 [c] ± 0.25	3.69 [b] ± 0.31

Results are displayed as Mean ± Standard Error. [a,b,c] Noteworthy differences ($p < 0.05$) are observed among the means within the same rows of distinct litters. DMA; malondialdehyde, GSH; reduced glutathione, CAT; catalase.

3.3. Proinflammatory Cytokines and Apoptotic Biomarkers

Oral administration of DEHP statistically elevated hepatic and renal proinflammatory levels; TNF-α (~112% and 177%), IL-1β (~280% and 87%), NF-κβ (~41% and 107%), with augmented the proapoptotic marker; caspases-3 (~118% and 200%), while reducing the anti-apoptotic index level; Bcl-2 (~52% and 41%) as compared to control group indicating inflammatory and apoptotic conditions observed in both tissues with the kidney is predominant. Lutein administration simultaneously with DEHP significantly has anti-inflammatory activity, decreasing the proinflammatory cytokines and proapoptotic biomarkers in hepato-renal tissues. It also enhanced the anti-apoptotic index in both tissues, as recorded in Figures 1 and 2.

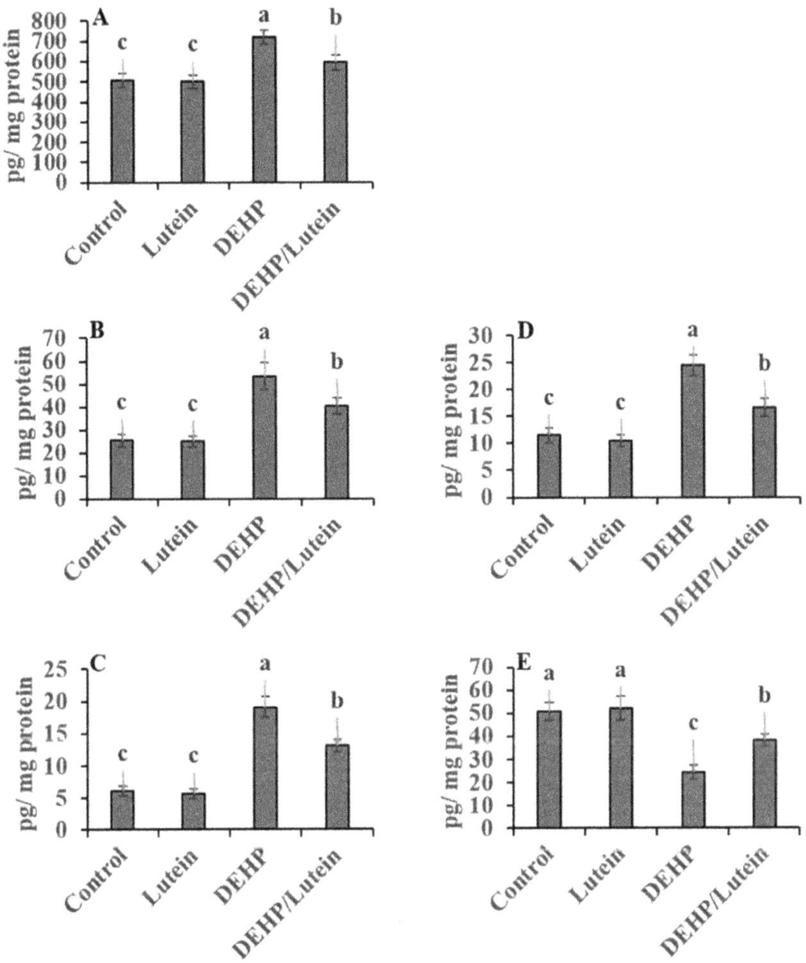

Figure 1. Hepatic proinflammatory and apoptotic biomarkers in DEHP and lutein-treated groups. (**A**); nuclear factor-κβ (NF-κβ), (**B**); tumor necrosis factor-α (TNF-α), (**C**); interleukin-1β (IL-1β), (**D**); caspase-3, (**E**); β cell lymphoma-2 (Bcl2). Results are displayed as Mean ± Standard Error. [a,b,c] Noteworthy differences ($p < 0.05$) are observed among the columns with distinct litters.

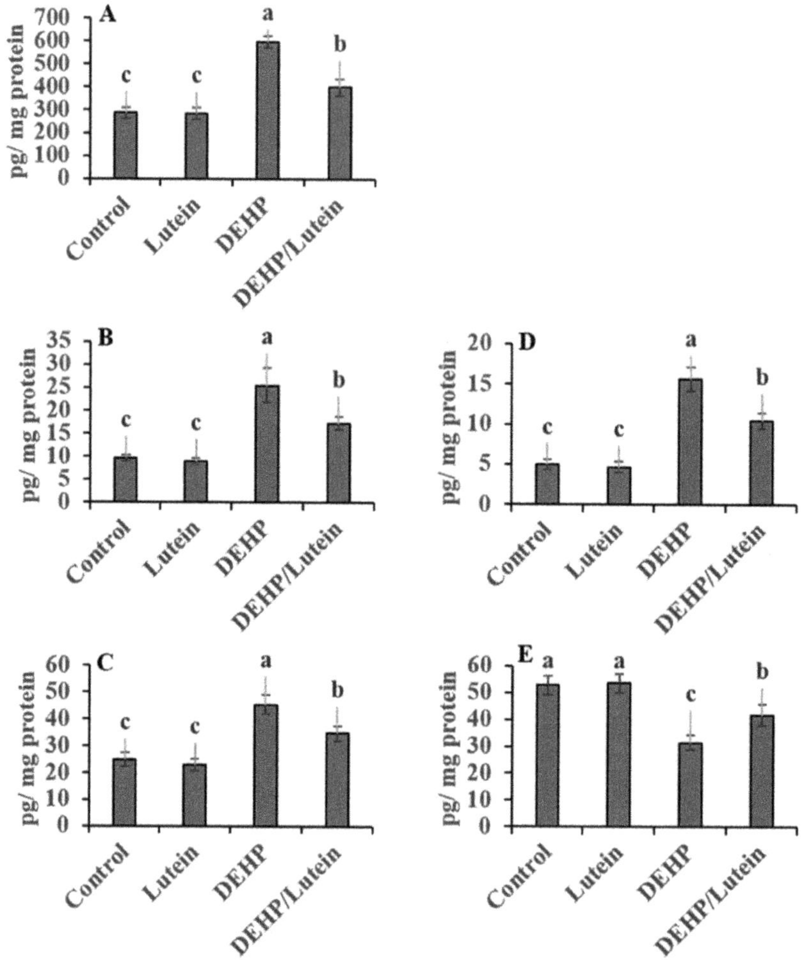

Figure 2. Renal proinflammatory and apoptotic biomarkers in DEHP and lutein-treated groups. (**A**); nuclear factor-κβ (NF-κβ), (**B**); tumor necrosis factor-α (TNF-α), (**C**); interleukin-1β (IL-1β), (**D**); caspase-3, (**E**); β cell lymphoma-2 (Bcl2). Results are displayed as Mean ± Standard Error. [a,b,c] Noteworthy differences ($p < 0.05$) are observed among the columns with distinct litters.

3.4. Histopathological Changes

3.4.1. Liver

Upon histopathological examination, the liver of the control group appeared with normal histoarchitectures (Figure 3a). In contrast, hepatic tissues of the DEHP-intoxicated group showed congestion of hepatic sinusoids with hemorrhage (Figure 3b), in addition to widespread hepatocytes hydropic degeneration and the presence of necrotic foci (Figure 3c). Lutein co-treated with DEHP showed only low-grade hydropic degeneration of hepatocytes (Figure 3d) and congestion of blood vessels, which was accompanied by perivascular inflammatory cell infiltration (Figure 3e).

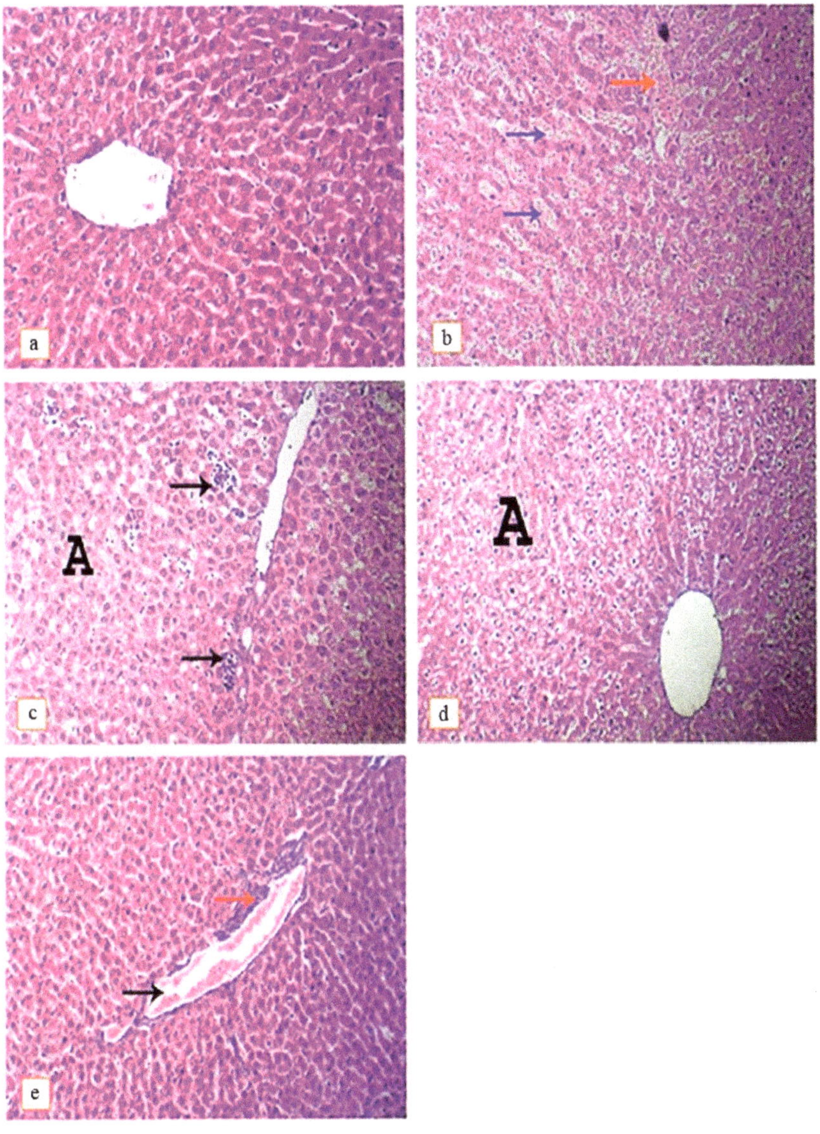

Figure 3. Photomicrograph of rat's liver intoxicated with DEHP and treated with lutein, H&E. (×400). (**a**); control group, showing normal histoarchitecture, (**b**); DEHP-intoxicated group, showing congestion of hepatic sinusoids (blue arrows), hemorrhage (red arrow), (**c**); DEHP-intoxicated group, showing multi-focal hepatic necrosis (arrows) and hydropic degeneration of hepatocytes (A), (**d**); DEHP and lutein co-treated group, showing hydropic degeneration of hepatocytes (A), (**e**); DEHP and lutein co-treated group, showing congestion of blood vessel (black arrow) and perivascular infiltration of mononuclear inflammatory cells (red arrow).

3.4.2. Kidney

Histopathological examination of renal tissues of the control group revealed the presence of normal glomeruli and renal tubules without any detected lesions (Figure 4b). On the contrary, DEHP-intoxicated rats showed the presence of interstitial inflammatory cells infiltration and tubular necrosis (Figure 4b), with vacuolar degeneration of tubular epithelium, presence of atrophied glomeruli, detached tubular epithelium and cystic dilatation of renal tubules (Figure 4c). On the other hand, administration of lutein with DEHP only causes congestion of interstitial blood vessels with mild focal tubular necrosis and cystic dilation (Figure 4d). The ameliorative effect of lutein administration against DEHP-induced hepato-nephrotoxicity-related lesions has been reflected in scores of these detected lesions, as shown in Table 3.

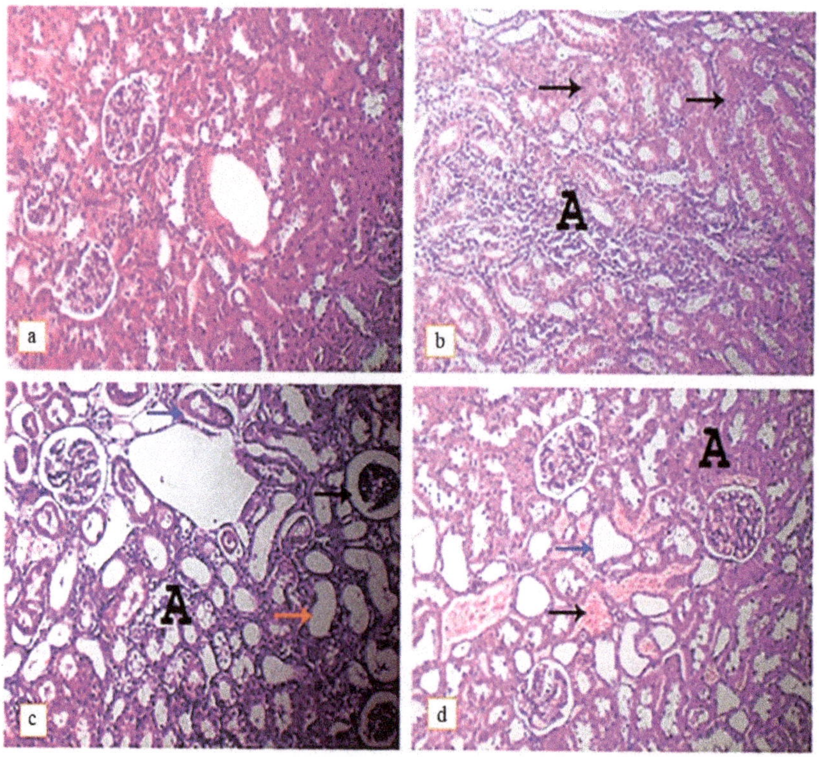

Figure 4. Photomicrograph of rat's kidneys intoxicated with DEHP and treated with lutein, H&E. (×400). (**a**); control group, showing normal histoarchitecture, (**b**); DEHP-intoxicated group, showing widespread interstitial mononuclear inflammatory cells infiltration (A) and tubular necrosis (arrows), (**c**); DEHP-intoxicated group, showing atrophied glomerulus (arrows), hydropic degeneration of tubular epithelium with interstitial inflammatory cells infiltration (A), cystic dilatation of renal tubules (red arrow) and presence of detached renal epithelium within the lumen of renal tubules (blue arrow), (**d**); DEHP and lutein co-treated group, showing congestion of interstitial blood vessels (black arrow) with focal necrosis of renal tubules (A) and cystic dilation of renal tubules.

Table 3. Semi-quantitative scoring results of the detected hepato-renal lesions.

	Incidence [1] and Severity [2] of Histopathological Lesions							
	DEHP Intoxicated Rats				**DEHP and Lutein-Treated Rats**			
	Absent (-)	**Mild (+)**	**Moderate (++)**	**Severe (+++)**	**Absent (−)**	**Mild (+)**	**Moderate (++)**	**Severe (+++)**
				Liver				
1-Hydropic degeneration	0	2	0	5	3	1	2	1
2-Congested sinusoids	2	2	2	1	5	2	0	0
3-Congestion of blood vessels	1	2	2	2	2	2	1	2
4-Perivascular infiltration of inflammatory cells	3	1	2	1	3	2	1	1
4-Necrotic foci	1	3	1	2	3	4	0	0
5-Hemorrhage	2	1	2	2	6	0	1	0
				Kidney				
1-Atrophied glomeruli	3	1	2	1	5	2	0	0
2-Congested blood vessels	2	2	1	2	4	1	1	1
2-Necrotic tubules	0	2	2	3	2	2	1	2
3-Interstitial nephritis	2	1	1	3	4	2	1	0
4-Hydropic degeneration of tubular epithelium	3	3	1	0	4	2	0	1
5-Cystic dilatation	1	3	1	2	3	1	1	2
6-Detached tubular epithelium	2	1	2	2	5	1	1	0

[1] Number of rats with lesions per total examined (7 rats). [2] Severity of lesions was graded by estimating the percentage area affected in the entire section.

4. Discussion

DEHP has attracted attention during the last few years as one of the most dangerous environmental toxicants due to its wide usage in the production of polyvinyl chloride (PVC)-based products such as food containers, water pipes, and medical devices which can be incriminated in human and animal toxicity [27]. Most environmental toxicant exerts a deleterious effect on the body's organs through the generation of ROS [28], which could disturb cell functions, causing cell death and even carcinogenesis [29]. Concerning DEHP-induced hepatotoxicity, several studies have indicated that such toxicity is strongly related to redox balance disturbance and depletion of tissue antioxidants [30]. In this context, the increase in serum activity of hepatocytic transaminase enzymes (AST and ALT) and cholestatic enzyme (GGT) confirmed the toxic effect of DEHP on the liver, as reported before by Erkekoglu et al. [31]; these enzymes are released from hepatocytes and biliary epithelium lining respectively upon their destruction which led to their increase in serum [32]. DEHP's carcinogenic effect has been studied extensively [33], and the liver is one of the most predilection sites for DEHP carcinogenicity [34]. AFP is one of the major tumor-associated proteins [35] produced by the liver, and its production is increased in response to carcinogenesis, especially hepatocellular carcinoma [36]. So, the detected increase in serum level of AFP upon exposure to DEHP may indicate its carcinogenic effect on the liver. DEHP nephrotoxicity has been reported and confirmed in animals and cellular models [37]. Similarly, DEHP-related nephrotoxicity may be owed to DEHP-induced ROS production [38]. So, the detected elevation in serum level of creatinine and cystatin-C (kidney functional biomarkers) would confirm DEHP's nephrotoxic effect as reported before [23]. Albumin is one of the negative acute phase proteins (which decrease in response to inflammation), and globulins are positive acute phase proteins (which increase in response to inflammation); acute phase proteins are increased or decreased due to the release of inflammatory cytokines [39], and this may explain the decrease in serum albumin level and increased serum globulins level in DEHP-intoxicated animals, in response to hepato-renal inflammatory state which will be discussed later. Also, defective hepatic albumin production or renal albumin loss due to DEHP-induced hepatorenal toxicity could be another explanation for decreased albumin levels [32]. In our study, the previously discussed DEHP-induced redox homeostasis disturbance in renal and hepatic tissues was reflected in the form of an elevation of MDA content in these tissues and depletion of their content of enzymatic and non-enzymatic defensive antioxidants (CAT, SOD, and GSH). Production of ROS may induce cell death due to the activation of different pathways and factors, including inflammatory and apoptotic factors [40]. The latter may explain the significant increase in the renal and hepatic levels of proinflammatory cytokines (IL-1β and TNF-α), as the increase in ROS production would enhance the inflammatory process and stimulate the release of proinflammatory cytokines. These cytokine releases could induce further ROS production [41]. Nf-κB plays a crucial role in the induction of inflammation, as it can increase the transcription of proinflammatory cytokines and chemokines [42]. Another theory for DEHP-induced toxicity includes increased production of Nf-κB in affected tissues [23], as reported in our study. Also, Huang et al. [43] reported upregulation of proinflammatory cytokines (NF-κβ, IL-6, IL-1β, and TNF-α) upon exposure of quail to DEHP. Fortunately, as a part of the anti-inflammatory activity of carotenoids, lutein's ability to reduce Nf-κB was proved before [44], so the level of inflammatory Nf-κB was decreased in hepatic and renal tissues of DEHP-intoxicated rats treated with lutein. Apoptosis is strongly implicated in DEHP-induced toxicity [45], which is mediated by several apoptotic proteins such as caspase-3, and this may explain the increase in hepatic and renal tissue content of caspase-3 (apoptotic protease). Also, the elevation in the caspase-3 level may be attributed to the decrease in Bcl-2 (anti-apoptotic protein), which is responsible for the regulation of the caspase-3 level [46]. Previously discussed biochemical changes related to DEHP hepato-renal toxicity were associated with variable lesions in both hepatic and renal tissues upon histopathological evaluation. Lutein antioxidant, anti-inflammatory, and anti-apoptotic activities were detected and proven previously [47]. So, based on these activities, its role in the amelioration of hepato-renal injuries and toxicity was proved in several studies [18]. Similarly, in our study, lutein successfully mitigated DEHP-induced hepato-renal toxicity and its related biochemical, inflammatory, and apoptotic biomarkers alterations; the lutein ameliorative effect was confirmed histo-pathologically through lowering the score of the induced lesions in hepatic and renal tissues. The structural base of the antioxidative effect of lutein is believed to contribute to the delocalization of unpaired electrons by its conjugated double-bonded structure. This allows lutein to effectively scavenge free radicals such as superoxide, hydroxyl, and peroxyl radicals. By scavenging these free radicals, lutein helps prevent the generation of oxidative stress, which can lead to cellular damage. This is why lutein is often considered a potent antioxidant with potential health benefits [48]. Finally, it could be concluded that lutein, as one of the natural ingredients of our daily food, can abrogate DEHP-induced hepato-nephrotoxicity due to its potent antioxidant, anti-inflammatory, and anti-apoptotic effects.

5. Conclusions

This study demonstrates that exposure to DEHP resulted in a significant increase in liver enzyme activity and alpha-fetoprotein, creatinine, cystatin-C, and disruptions in serum protein profile. Furthermore, DEHP intoxication disrupted the redox balance in hepato-renal tissues, leading to an increase in proinflammatory cytokines, nuclear factor kappa B (Nf-κB), and apoptotic marker (caspase-3) while decreasing the level of anti-apoptotic protein (Bcl-2) in these tissues. However, administering lutein at a dose of 40 mg/kg b.w efficiently facilitated the changes in serum biochemical constituents, hepato-renal oxidative disturbance, inflammatory and apoptotic markers, and histopathological alterations induced by DEHP intoxication. As a natural carotenoid, lutein is likely to protect against DEHP toxicity.

Author Contributions: D.R.S.G.E.-K., M.A.L. and B.S.A.: Supervised, designed, performed experiments, and revised the manuscript. A.F.E.-k., H.I.G. and M.S.: Performed experiments, analyzed data, wrote and finalized the manuscript. D.R.S.G.E.-K., M.A.L. and B.S.A.: Performed experiments. D.R.S.G.E.-K. and M.A.L.: Performed experiments, wrote, interpreted data, and finalized the manuscript. All authors have read and agreed to the published version of the manuscript.

Funding: The authors thank the Deanship of Scientific Research at King Khalid University for funding this work through a large group Research Project under grant number RGP2/435/44. We appreciate the resources the Princess Nourah bint Abdulrahman University Researchers Supporting Project number (PNURSP2023R73), Princess Nourah bint Abdulrahman University, Riyadh, Saudi Arabia.

Institutional Review Board Statement: The Animal Care and Use Committee of the Faculty of Veterinary Medicine at the University of Alexandria in Egypt has approved animal care and use guidelines. code "Au-0.13-2022-12-12-177".

Informed Consent Statement: Not applicable.

Data Availability Statement: Upon request.

Acknowledgments: The authors thank the Deanship of Scientific Research at King Khalid University for funding this work through a large group Research Project under grant number RGP2/435/44. This work was supported by Princess Nourah bint Abdulrahman University Researchers Supporting Project number (PNURSP2023R73), Princess Nourah bint Abdulrahman University, Riyadh, Saudi Arabia.

Conflicts of Interest: The authors declare no conflict of interest.

References

1. Chou, K.; Wright, R.O. Phthalates in food and medical devices. *J. Med. Toxicol.* **2006**, *2*, 126–135.
2. Koniecki, D.; Wang, R.; Moody, R.P.; Zhu, J. Phthalates in cosmetic and personal care products: Concentrations and possible dermal exposure. *Environ. Res.* **2011**, *111*, 329–336.
3. Latini, G. Monitoring phthalate exposure in humans. *Clin. Chim. Acta* **2005**, *361*, 20–29. [PubMed]
4. Fromme, H.; Gruber, L.; Schlummer, M.; Wolz, G.; Böhmer, S.; Angerer, J.; Mayer, R.; Liebl, B.; Bolte, G. Intake of phthalates and di(2-ethylhexyl) adipate: Results of the Integrated Exposure Assessment Survey based on duplicate diet samples and biomonitoring data. *Environ. Int.* **2007**, *33*, 1012–1020. [PubMed]
5. Chiellini, F.; Ferri, M.; Latini, G. Physical–chemical assessment of di-(2-ethylhexyl)-phthalate leakage from poly(vinyl chloride) endotracheal tubes after application in high risk newborns. *Int. J. Pharm.* **2011**, *409*, 57–61.
6. Tickner, J.A.; Schettler, T.; Guidotti, T.; McCally, M.; Rossi, M. Health risks posed by use of Di-2-ethylhexyl phthalate (DEHP) in PVC medical devices: A critical review. *Am. J. Ind. Med.* **2001**, *39*, 100–111.
7. Heudorf, U.; Mersch-Sundermann, V.; Angerer, J. Phthalates: Toxicology and exposure. *Int. J. Hyg. Environ. Health* **2007**, *210*, 623–634.
8. Erkekoglu, P.; Zeybek, N.D.; Giray, B.; Asan, E.; Arnaud, J.; Hincal, F. Reproductive toxicity of di(2-ethylhexyl) phthalate in selenium-supplemented and selenium-deficient rats. *Drug Chem. Toxicol.* **2011**, *34*, 379–389. [CrossRef]
9. Somasundaram, D.; Manokaran, K.; Selvanesan, B.; Bhaskaran, R. Impact of di-(2-ethylhexyl) phthalate on the uterus of adult Wistar rats. *Hum. Exp. Toxicol.* **2017**, *36*, 565–572. [CrossRef]
10. Kasahara, E.; Sato, E.F.; Miyoshi, M.; Konaka, R.; Hiramoto, K.; Sasaki, J.; Tokuda, M.; Nakano, Y.; Inoue, M. Role of oxidative stress in germ cell apoptosis induced by di(2-ethylhexyl)phthalate. *Biochem. J.* **2002**, *365*, 849–856.
11. Huang, Y.; Wu, C.; Ye, Y.; Zeng, J.; Zhu, J.; Li, Y.; Wang, W.; Zhang, W.; Chen, Y.; Xie, H. The increase of ROS caused by the interference of DEHP with JNK/p38/p53 pathway as the reason for hepatotoxicity. *Int. J. Environ. Res. Public Health* **2019**, *16*, 356. [CrossRef] [PubMed]
12. Zhang, W.; Shen, X.-Y.; Zhang, W.-W.; Chen, H.; Xu, W.-P.; Wei, W. The effects of di 2-ethyl hexyl phthalate (DEHP) on cellular lipid accumulation in HepG2 cells and its potential mechanisms in the molecular level. *Toxicol. Mech. Methods* **2017**, *27*, 245–252. [CrossRef]
13. Gasparovic, A.C.; Jaganjac, M.; Mihaljevic, B.; Sunjic, S.B.; Zarkovic, N. Assays for the measurement of lipid peroxidation. In *Cell Senescence: Methods and Protocols*; Springer: Berlin/Heidelberg, Germany, 2013; pp. 283–296.
14. Fuad, N.I.N.; Sekar, M.; Gan, S.H.; Lum, P.T.; Vaijanathappa, J.; Ravi, S. Lutein: A comprehensive review on its chemical, biological activities and therapeutic potentials. *Pharmacogn. J.* **2020**, *12*, 1769–1778. [CrossRef]

15. Sindhu, E.R.; Preethi, K.C.; Kuttan, R. Antioxidant Activity of Carotenoid Lutein In Vitro and In Vivo. 2010. Available online: http://nopr.niscpr.res.in/handle/123456789/9991 (accessed on 27 August 2023).
16. Kim, J.-H.; Na, H.-J.; Kim, C.-K.; Kim, J.-Y.; Ha, K.-S.; Lee, H.; Chung, H.-T.; Kwon, H.J.; Kwon, Y.-G.; Kim, Y.-M. The non-provitamin A carotenoid, lutein, inhibits NF-κB-dependent gene expression through redox-based regulation of the phosphatidylinositol 3-kinase/PTEN/Akt and NF-κB-inducing kinase pathways: Role of H_2O_2 in NF-κB activation. *Free Radic. Biol. Med.* **2008**, *45*, 885–896. [CrossRef] [PubMed]
17. Li, S.; Ding, Y.; Niu, Q.; Xu, S.; Pang, L.; Ma, R.; Jing, M.; Feng, G.; Tang, J.X.; Zhang, Q.; et al. Lutein has a protective effect on hepatotoxicity induced by arsenic via Nrf2 signaling. *BioMed Res. Int.* **2015**, *2015*, 315205. [CrossRef]
18. Bilgiç, S.; Gür, F.M.; Aktaş, İ. Biochemical and Histopathological Investigation of the Protective Effect of Lutein in Rat Kidney Exposed to Cisplatin. *Med. Rec.* **2022**, *4*, 433–438. [CrossRef]
19. Ouyang, B.; Li, Z.; Ji, X.; Huang, J.; Zhang, H.; Jiang, C. The protective role of lutein on isoproterenol-induced cardiac failure rat model through improving cardiac morphology, antioxidant status via positively regulating Nrf2/HO-1 signalling pathway. *Pharm. Biol.* **2019**, *57*, 529–535. [CrossRef]
20. Zhang, W.L.; Zhao, Y.N.; Shi, Z.Z.; Cong, D.; Bai, Y.S. Lutein inhibits cell growth and activates apoptosis via the PI3K/AKT/mTOR signaling pathway in A549 human non-small-cell lung cancer cells. *J. Environ. Pathol. Toxicol. Oncol.* **2018**, *37*, 341–350. [CrossRef]
21. Ashari, S.; Naghsh, N.; Salari, Y.; Barghi, N.G.; Bagheri, A. Dimethyl Fumarate Attenuates Di-(2-Ethylhexyl) Phthalate-Induced Nephrotoxicity Through the Nrf2/HO-1 and NF-κB Signaling Pathways. *Inflammation* **2023**, *46*, 453–467. [CrossRef]
22. Aydemir, D.; Karabulut, G.; Gok, M.; Barlas, N.; Ulusu, N.N. Data the DEHP induced changes on the trace element and mineral levels in the brain and testis tissues of rats. *Data Brief* **2019**, *26*, 104526. [CrossRef]
23. Ashari, S.; Karami, M.; Shokrzadeh, M.; Bagheri, A.; Ghandadi, M.; Ranaee, M.; Dashti, A.; Mohammadi, H. Quercetin ameliorates Di(2-ethylhexyl) phthalate-induced nephrotoxicity by inhibiting NF-κB signaling pathway. *Toxicol. Res.* **2022**, *11*, 272–285. [CrossRef] [PubMed]
24. Gao, H.-T.; Shi, H.-Y.; Dai, Q.-M.; Li, A.-Q.; Yang, L.; Sun, Y.; Jin, S.-Y.; Xia, L.-Z. Glycerin monostearate aggravates male reproductive toxicity caused by di(2-ethylhexyl) phthalate in rats. *Curr. Med. Sci.* **2019**, *39*, 1003–1008. [CrossRef] [PubMed]
25. Du, S.-Y.; Zhang, Y.-L.; Bai, R.-X.; Ai, Z.-L.; Xie, B.-S.; Yang, H.-Y. Lutein prevents alcohol-induced liver disease in rats by modulating oxidative stress and inflammation. *Int. J. Clin. Exp. Med.* **2015**, *8*, 8785. [PubMed]
26. Banchroft, J.; Stevens, A.; Turner, D. *Theory and Practice of Histological Techniques Fourth Ed Churchil Livingstone*; Elsevier health sciences: Amsterdam, The Netherlands, 1996.
27. Akingbemi, B.T.; Ge, R.; Klinefelter, G.R.; Zirkin, B.R.; Hardy, M.P. Phthalate-induced Leydig cell hyperplasia is associated with multiple endocrine disturbances. *Proc. Natl. Acad. Sci. USA* **2004**, *101*, 775–780. [CrossRef]
28. Oberley, T.D.; Zhong, W.; Szweda, L.I.; Oberley, L.W. Localization of antioxidant enzymes and oxidative damage products in normal and malignant prostate epithelium. *Prostate* **2000**, *44*, 144–155. [CrossRef] [PubMed]
29. Nose, K. Role of reactive oxygen species in the regulation of physiological functions. *Biol. Pharm. Bull.* **2000**, *23*, 897–903. [CrossRef]
30. Chen, H.; Zhang, W.; Rui, B.B.; Yang, S.M.; Xu, W.P.; Wei, W. Di(2-ethylhexyl) phthalate exacerbates non-alcoholic fatty liver in rats and its potential mechanisms. *Environ. Toxicol. Pharmacol.* **2016**, *42*, 38–44. [CrossRef] [PubMed]
31. Erkekoglu, P.; Zeybek, N.D.; Giray, B.K.; Rachidi, W.; Kızılgün, M.; Hininger-Favier, I.; Favier, A.; Asan, E.; Hincal, F. The effects of di(2-ethylhexyl) phthalate on rat liver in relation to selenium status. *Int. J. Exp. Pathol.* **2014**, *95*, 64–77. [CrossRef]
32. Latimer, K.S.; Mahaffey, E.A.; Prasse, K.W. *Veterinary Laboratory Medicine: Clinical Pathology*; Iowa State Press: Story County, IA, USA, 2003.
33. Ito, Y.; Nakajima, T. PPARα-and DEHP-induced cancers. *PPAR Res.* **2008**, *2008*, 759716. [CrossRef]
34. Voss, C.; Zerban, H.; Bannasch, P.; Berger, M.R. Lifelong exposure to di-(2-ethylhexyl)-phthalate induces tumors in liver and testes of Sprague–Dawley rats. *Toxicology* **2005**, *206*, 359–371. [CrossRef]
35. Terentiev, A.; Moldogazieva, N. Alpha-fetoprotein: A renaissance. *Tumor Biol.* **2013**, *34*, 2075–2091. [CrossRef]
36. Galle, P.R.; Foerster, F.; Kudo, M.; Chan, S.L.; Llovet, J.M.; Qin, S.; Schelman, W.R.; Chintharlapalli, S.; Abada, P.B.; Sherman, M.; et al. Biology and significance of alpha-fetoprotein in hepatocellular carcinoma. *Liver Int.* **2019**, *39*, 2214–2229. [CrossRef]
37. Wu, C.-T.; Wang, C.-C.; Huang, L.-C.; Liu, S.-H.; Chiang, C.-K. Plasticizer di-(2-ethylhexyl) phthalate induces epithelial-to-mesenchymal transition and renal fibrosis in vitro and in vivo. *Toxicol. Sci.* **2018**, *164*, 363–374. [CrossRef]
38. Ashari, S.; Karami, M.; Shokrzadeh, M.; Ghandadi, M.; Ghassemi-Barghi, N.; Dashti, A.; Ranaee, M.; Mohammadi, H. The implication of mitochondrial dysfunction and mitochondrial oxidative damage in di(2-ethylhexyl) phthalate induced nephrotoxicity in both in vivo and in vitro models. *Toxicol. Mech. Methods* **2020**, *30*, 427–437. [CrossRef]
39. Gabay, C.; Kushner, I. Acute-phase proteins and other systemic responses to inflammation. *New Engl. J. Med.* **1999**, *340*, 448–454. [CrossRef] [PubMed]
40. Redza-Dutordoir, M.; Averill-Bates, D.A. Activation of apoptosis signalling pathways by reactive oxygen species. *Biochim. Et Biophys. Acta (BBA)-Mol. Cell Res.* **2016**, *1863*, 2977–2992. [CrossRef]
41. Elmarakby, A.A.; Sullivan, J.C. Relationship between oxidative stress and inflammatory cytokines in diabetic nephropathy. *Cardiovasc. Ther.* **2012**, *30*, 49–59. [CrossRef] [PubMed]

42. Lawrence, T. The nuclear factor NF-κB pathway in inflammation. *Cold Spring Harb. Perspect. Biol.* **2009**, *1*, a001651. [CrossRef] [PubMed]
43. Huang, Y.-Q.; Tang, Y.-X.; Qiu, B.-H.; Talukder, M.; Li, X.-N.; Li, J.-L. Di-2-ethylhexyl phthalate (DEHP) induced lipid metabolism disorder in liver via activating the LXR/SREBP-1c/PPARα/γ and NF-κB signaling pathway. *Food Chem. Toxicol.* **2022**, *165*, 113119. [CrossRef]
44. Armoza, A.; Haim, Y.; Basiri, A.; Wolak, T.; Paran, E. Tomato extract and the carotenoids lycopene and lutein improve endothelial function and attenuate inflammatory NF-κB signaling in endothelial cells. *J. Hypertens.* **2013**, *31*, 521–529. [CrossRef]
45. Amara, I.; Ontario, M.L.; Scuto, M.; Lo Dico, G.M.; Sciuto, S.; Greco, V.; Abid-Essefi, S.; Signorile, A.; Salinaro, A.T.; Calabrese, V. Moringa oleifera protects SH-SY5YCells from DEHP-induced endoplasmic reticulum stress and apoptosis. *Antioxidants* **2021**, *10*, 532. [CrossRef] [PubMed]
46. Swanton, E.; Savory, P.; Cosulich, S.; Clarke, P.; Woodman, P. Bcl-2 regulates a caspase-3/caspase-2 apoptotic cascade in cytosolic extracts. *Oncogene* **1999**, *18*, 1781–1787. [CrossRef] [PubMed]
47. Gündoğdu, B.; Taş, H.G.; Süleyman, B.; Mamedov, R.; Yüce, N.; Kuyrukluyildiz, U.; Süleyman, H. Effect of lutein on oxidants and proinflammatory cytokine-related liver ischemia-reperfusion injury. *Acta Pol. Pharm.* **2022**, *79*, 129–135. [CrossRef]
48. Singh, J.; Upadhyay, A.; Bahadur, A.; Singh, B.; Singh, K.; Rai, M. Antioxidant phytochemicals in cabbage (*Brassica oleracea* L. var. capitata). *Sci. Hortic.* **2006**, *108*, 233–237. [CrossRef]

toxics

MDPI

Article

Combined Toxicities of Di-Butyl Phthalate and Polyethylene Terephthalate to Zebrafish Embryos

Qiang Zhang [1,2], Wenjie Ma [2] and Jingmin Zhu [1,2,*]

[1] School of Fishery, Zhejiang Ocean University, Zhoushan 316022, China
[2] Guangxi Key Laboratory of Marine Environmental Change and Disaster in the Beibu Gulf, Beibu Gulf University, Qinzhou 535011, China
[*] Correspondence: jmzhu@zjou.edu.cn

Abstract: The increasing concern for the ecological risks of microplastics (MPs) as carriers of hydrophobic organic contaminants is evident. Di-butyl phthalate (DBP) is extensively utilized as an additive in plastic products, and both DBP and MPs are widespread in the environment. However, the combined toxicity of these substances remains uncertain. In this study, zebrafish embryos were employed to assess the toxic effects of polyethylene terephthalate (PET, MPs) and DBP, with a focus on the DBP toxicities influenced by PET. The embryonic chorion was partially covered by PET particles, and PET led to a delayed hatching of zebrafish embryos without inducing death or teratogenesis. On the other hand, exposure to DBP considerably inhibited the hatching of embryos, leading to severe lethal and teratogenic effects. The most common phenotypes induced by DBP exposure were delayed yolk sac absorption and pericardial edema. The mortality increased in co-treatment with 100 particles/mL PET and 2 mg/L DBP at 24 hpf and 48 hpf. The malformation phenotype, bent notochord, and delayed yolk sac absorption became more severe in 1 mg/L DBP exposition with the co-exposure of 100 particles/mL PET at 72 hpf. PET might act as a carrier that enhances the bioavailability of ambient DBP.

Keywords: polyethylene terephthalate; di-butyl phthalate; zebrafish embryos; malformation phenotype; transport carrier

1. Introduction

Plastics are widely used in people's daily lives and are easily discarded. Plastics with diameters between 1 μm and 5 mm are described as microplastics (MPs), which are present in marine systems worldwide and especially concentrated in estuaries, lakes, and coastal waters where humans are abundant [1]. The amount of MPs and plastic debris in the global ocean was estimated to be at least 270,000 tons in 2014 [2], and the most common types were polystyrene (PS), polyethylene terephthalate (PET), and polypropylene (PP) [3]. PET is widely used as synthetic fibers, and the annual global production was approximately 53.3 million tons in 2016 [4,5]. MPs can be ingested by various organisms and are known for vector transport of hydrophobic organic contaminants, and there are growing concerns regarding their potential adverse effects on ecosystems and human health [6–10].

It was reported that MPs could affect photosynthesis and metabolism in phytoplankton [11]. A study by Wu et al. found that MPs could activate the expression of antibiotic-resistance genes and promote the spread of pathogens in microorganisms [12]. MPs can be enriched in the gill, stomach, and gut of organisms, which could be passed along the food chain [13] and cause inflammation [13]. In zebrafish, the toxicities of MPs include developmental delay [14,15], intestinal and metabolism damage [16–18], oxidative stress [15,18–20], immunotoxicity [15,21], neurotoxicity and locomotor toxicity [15,22,23], genotoxicity [16,22], and reproductive toxicity [22,24]. Studies have also found the effects of microplastics were often more severe during early development. Cormier et al. (2022)

found that the exposure of early life stages to particles in water induced a decrease in larval swimming activity [25]. Chronic trophic exposure to MPs reduced growth and reproduction for both fish F0 and survival, growth, and behavior of F1 larval offspring were affected by MPs [25]. Rochman et al. (2014) observed 20 μm PS and PE microplastics present in zebrafish eleutheroembryos (from the larval stage to the onset of active feeding) at concentrations as low as 0.2 mg/L caused growth inhibition and delayed inflating of swim bladders, or their absence [26]. The ingestion of plastic debris at environmentally relevant concentrations may alter endocrine system function in adult fish and warrants further research [26]. Although many studies have reported the toxicity of microplastics in the lab, the actual environmental risks of microplastics and their associated chemicals remain largely unknown [27,28].

Phthalates (PAEs) are used as additives in plastic products to increase elasticity and flexibility and also as carriers of pesticides and insect repellents [29–32]. PAEs are not chemically bonded to plastic polymers and can easily migrate from products to the environment, having become common in the environment [33,34]. PAEs are not easily degraded and can be enriched in biological tissues [35]. Six PAEs, including dioctyl phthalate (DEHP), diethyl phthalate (DEP), dimethyl phthalate (DMP), benzyl butyl phthalate (BBP), di-n-octyl phthalate (DNOP), and di-butyl phthalate (DBP), have been listed as priority pollutants by the United States Environmental Protection Agency (USERA) [36]. DBP is one of the most commonly used plasticizers and is widely used in children's toys, plastic food containers, cosmetics, pharmaceuticals, and insect repellents [37]. DBP has been frequently detected in surface waters, wastewater, sewage sludge, sediments, and aquatic organisms worldwide at ng/L to μg/L levels [38–42]. The maximum value of DBP (3.55 μg/L) in a water sample from Yangtze River Delta, China, exceeds the limit value, which implies that there is a potential impact on the environment or human body [43]. The concentrations of DBP in Hangzhou Bay (2.85–18.0 μg/L) and Zhenjiang (0.330–12.6 μg/L) were much higher than those reported in the water in the Jiangsu section in Yangtze River (0.105–0.286 μg/L), the East China Sea (0.088–4.96 μg/L), Taihu Lake (nd–2.54 μg/L), and Yangtze River Delta (nd–7.19 μg/L) [43]. Moreover, the average concentration of phthalate esters (PAEs) in sediment samples was found to be around 1200 μg/kg, which was about 300 times higher than the concentration in surface water samples (4.11 μg/kg). Lee et al. (2019) reported a positive correlation between the concentrations of phthalate esters (PAEs) in sediment and their log Kow values, suggesting that PAEs with higher Kow values have a greater tendency to adsorb to sediment [44]. DBP poses risks to aquatic organisms, even at low levels. DBP can bioaccumulate in the food chain and biomagnify to high levels that threaten fish-eating wildlife and humans. It was reported that exposure to DBP had resulted in yolk sac abnormalities, skeletal defects, spinal curvatures, abnormal movement, craniofacial defects, cardiac defects, defects in eye vascularization, as well as immunotoxicity in zebrafish embryos or larvae [33,45–49].

Due to its hydrophobic properties and large surface area, MPs possess strong adsorption affinity to environmental pollutants, such as POPs and heavy metals [50–52]. MPs could increase the neurotoxicity of bisphenol A to zebrafish [53]. A study by Zhang et al. found that the mixture of MPs and Cd resulted in antagonistic toxicity under low concentration of MPs (0.05, 0.1 mg/L), while there was synergistic sublethal toxicity under high levels of MPs (1, 5, 10 mg/L) on zebrafish embryos [3]. Both MPs and DBP are common in the environment, but their combined toxicity is still unclear. Zebrafish is one of the most widely used model species to study the developmental toxicity of chemicals. Zebrafish embryos are more sensitive to environmental pollutants than adult fishes [54–56], and the fertilization is external, facilitating toxicant exposure at defined concentrations [57]. The present study aimed to investigate the toxic effects of the combination of polyethylene terephthalate (PET) and di-butyl phthalate (DBP) on zebrafish embryos, with a focus on the potential influence of PET on DBP toxicity.

2. Materials and Methods

2.1. Chemicals and Reagents

In this study, di-butyl phthalate (DBP) with a purity of 99.7%, tricaine (MS-222) used for anaesthesia, and Dimethyl sulfoxide (DMSO) were procured from Aladdin (Shanghai, China). Colorless polyethylene terephthalate (PET) particles were obtained from the local market and simultaneously sold online at https://m.tb.cn/h.UsDYlDY?tk=ksf3dk4Sw9X (18 May 2023). To aid observation, the PET particles were stained bright red using Nile red (CAS 7385-67-3, Aladdin, Bay City, MI, USA) in the laboratory. Steel sieves with a pore size of 150 μm and 100 μm were used to screen out particles with sizes between 100 μm and 150 μm. Before the exposure experiments, the prepared PET particles were visually inspected and validated through photographs. This approach was taken to ensure that only high-quality particles were used, and to minimize the possibility of any contamination or variability in the results.

2.2. Zebrafish Maintenance and Embryo Collection

The zebrafish used in this study were kept in a temperature-controlled room at 28 ± 0.5 °C with a 10:14 h dark:light cycle in a closed flow-through system that utilized charcoal-filtered tap water. The experimental procedures followed the OECD guidelines for chemical testing [58]. The fish embryos were collected and examined under a stereo microscope at 4 h post-fertilization (hpf) to ensure their health and viability. We individually weighed 0.294 g of $CaCl_2 \cdot 2H_2O$, 0.123 g of $MgSO_4 \cdot 7H_2O$, 0.065 g of $NaHCO_3$, and 0.006 g of KCl and dissolved them in 1 liter of fully aerated deionized purified water. After filtration, the solution was used for embryo culture. This ensured that the fish were kept in a healthy and optimal environment throughout the course of the study.

2.3. Acute Exposure Experiments of PET and DBP

In the DBP exposure experiment, the low concentration group was set at 0.05 mg/L, which was based on the environmental concentration (~20 μg/L, China) [19]. In the PET exposure experiment, the low concentration group was set at 1 particle/mL. The microplastic pollution in aquaculture water is about 50 particles per liter [59]. Considering the uneven distribution of microplastics in the exposure solution, the exposure concentration was set higher than the environmental concentration. The concentration gradient was set by increasing 1.5–5 times based on the lowest concentration, and a total of 5 exposure concentrations were established. DBP was dissolved in DMSO and then diluted into the embryo culture medium. The medium was filtered through a 0.45 μm filter membrane and maintained at 28 °C for the embryo exposure experiments. Embryos were exposed to 0.1% DMSO and 0.05, 0.2, 1, 2, and 3 mg/L DBP. PET stocks with embryo culture medium were sonicated at 40 kHz for 1 min prior to quantification and use for exposures. The exposure concentration gradient of PET was set at 0, 1, 10, 50, 100, and 200 particles/mL. Exactly 10 mL of suspension was added into a glass Petri dish with 6 cm diameter, and then 20 fertilized zebrafish embryos (screened at 4 hpf) were exposed in each dish and maintained in an incubator at 28 ± 0.5 °C until 96 hpf. Each concentration was tested in quadruplicate, with 20 embryos per replicate, resulting in a total of 80 embryos per concentration. (n = 4, 80 individuals). Embryos were observed at 12 hpf, 24 hpf, 48 hpf, 72 hpf, and 96 hpf, and the exposure suspension was changed every 24 h.

2.4. Measurement of the Acute Toxicity Indicators

The measurement of the acute toxicity indicators was recorded with reference to Cheng et al. [50]. The acute toxicity indicators included survival, hatching rate, and morphological malformations. Hatching success and survival rates are typically measured at different time points after exposure, and statistical analyses are performed to determine if there are significant differences between exposed and control groups. Morphological abnormalities are evaluated through visual inspection and imaging techniques. In this study, the mortality rate was recorded at 24 hpf, 48 hpf, 72 hpf, and 96 hpf, and the hatching rate was recorded at 48 hpf and 72 hpf. The malformation phenotypes, bent notochord, delayed yolk sac absorption, pericardial edema, and small eyes were analyzed using the image processing software Adobe Photoshop CS4 (CA, USA). After the measurement, the embryos or larvae were returned to the dish for subsequent experiments.

2.5. Statistical Analysis

After collecting data from the experiments, statistical analysis was performed using three different software: IBM SPSS 26 Statistics (IBM Corp., Armonk, NY, USA), GraphPad Prism 5.0 (GraphPad Software Inc., San Diego, CA, USA), and Origin 9.0 (OriginLab Corporation, Northampton, MA, USA). The first step in the analysis was to test for the homogeneity of variance, which is the assumption that the variances of the groups being compared are equal. This was done using Levene's statistic, and the homogeneity value was considered to be greater than 0.05. The mean differences between the control and DBP/PET exposures were then evaluated using one-way analysis of variance (ANOVA) with Dunnett post-hoc test. ANOVA is a widely used statistical method to compare means of three or more groups, and the post-hoc test is used to determine which groups differ significantly from each other. For the comparison of two groups, the independent sample *t*-test was used. This statistical method tests the difference between the means of two independent groups. In the figures, the letters above the bars indicate significant differences, with $p < 0.05$ considered statistically significant. If two groups have the same letter, then they are not significantly different from each other. To ensure the reliability and reproducibility of the results, the experiments were repeated four times. The results were expressed as mean ± SD, which is a standard way of presenting statistical data.

3. Results

3.1. PET Particles Made for Exposure Experiment

The prepared PET particles were examined using a Cnoptec SZ680 stereomicroscope (Chongqing, China), and images were captured using an AxioCam digital camera (Figure 1A). The particles were characterized using a micro-Fourier transformed infrared spectroscope (μ-FT-IR, Nicolet iN10 MX, Thermo Fisher Scientific, Waltham, MA, USA) in transmittance mode. The obtained spectrum was compared with the library of polymers provided by Thermo Fisher Scientific in their OMNIC Picta 1 software (Waltham, MA, USA), with a match quality index of >90% (Figure 1B). The images revealed that PET particles could attach to the zebrafish embryonic chorion before the embryo hatched (Figure 1C).

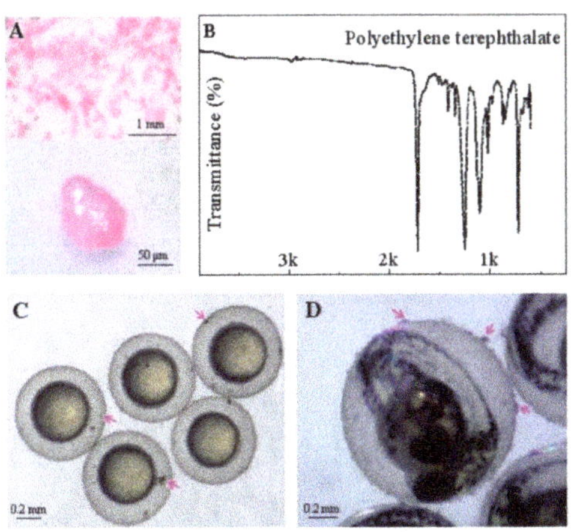

Figure 1. The characteristics of PET particles in this study. (**A**) Morphotype and size distribution of PET; (**B**) Transmittance spectrum analysis of PET; (**C**,**D**) PET particles (arrows) adsorbed on zebrafish embryonic chorion at 6 hpf (**C**) and 48 hpf (**D**).

3.2. Single Exposure of PET and DBP

Zebrafish embryos usually hatch into larvae during 48–96 hpf and mostly hatch at 72 hpf [60]. PET exposure at 50–200 particles/mL significantly delayed the hatching of zebrafish embryos at 48 hpf, while no significant difference was observed at 72 hpf (Figure 2A). A mortality rate of 6–9% was observed in the 200 particles/mL PET group at 72 hpf and 96 hpf (Figure 2B). After being treated with DBP, the hatching rates of embryos with all concentrations were significantly lower than the blank control group at 48 hpf, and inhibition was also observed at 72 hpf with 1–3 mg/L DBP expositions (Figure 2C). DBP exposure at 2–3 mg/L increased the mortality of zebrafish embryos during 24–96 hpf, especially at 72 hpf and 96 hpf, where the mortality reached 100% (Figure 2D).

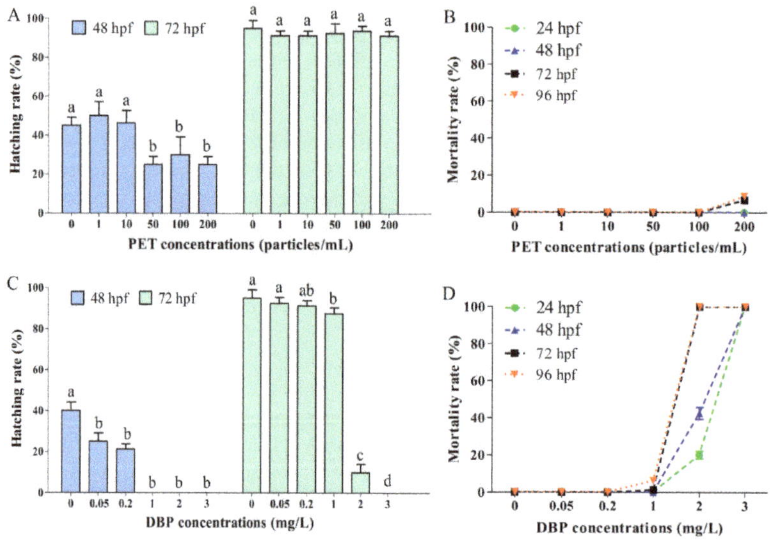

Figure 2. Hatching and mortality rates of zebrafish embryos under PET and DBP expositions. (**A**) Hatching rate at 48 hpf and 72 hpf with PET exposition; (**B**) Mortality rate at 24 hpf, 48 hpf, 72 hpf, and 96 hpf with PET exposition; (**C**) Hatching rate at 48 hpf and 72 hpf with DBP exposition; (**D**) Mortality rate at 24 hpf, 48 hpf, 72 hpf, and 96 hpf with DBP exposition; values represent mean ± SD ($n = 4$); the letters above the bars indicate significant differences ($p < 0.05$). If two arbitrary groups have the same letter, then they are not significantly different.

No obvious toxic effects on deformity were observed after PET exposure. Exposure to DBP induced severe malformations in zebrafish embryos, including short body length, delayed yolk sac absorption, pericardial edema, bent notochord, and small eyes (Figure 3A). The body length decreased by 2–16% in groups treated with 0.2–1 mg/L DBP compared with the blank control group (Figure 3B). The severity and malformation rate of delayed yolk sac absorption and pericardial edema malformations increased in a DBP concentration-dependent manner (Figure 3C,D).

3.3. DBP Toxicities Affected by PET

A reduction in hatching rate of 10–15% was observed in the 10–100 particles/mL PET groups (Figure 4A). Less than 1% of embryos in the 0.2–2 mg/L DBP groups hatched at 48 hpf, while hatching increased after the addition of PET (Figure 4A). Compared with single DBP exposure, the hatching rate at 72 hpf did not change under DBP + PET exposition (Figure 4B). The mortality increased in co-treatment with 100 particles/mL PET and 2 mg/L DBP at 24 hpf and 48 hpf only (Figure 4C). At 96 hpf, no influence of PET treatment was shown, considering the 100% mortality produced by DBP alone.

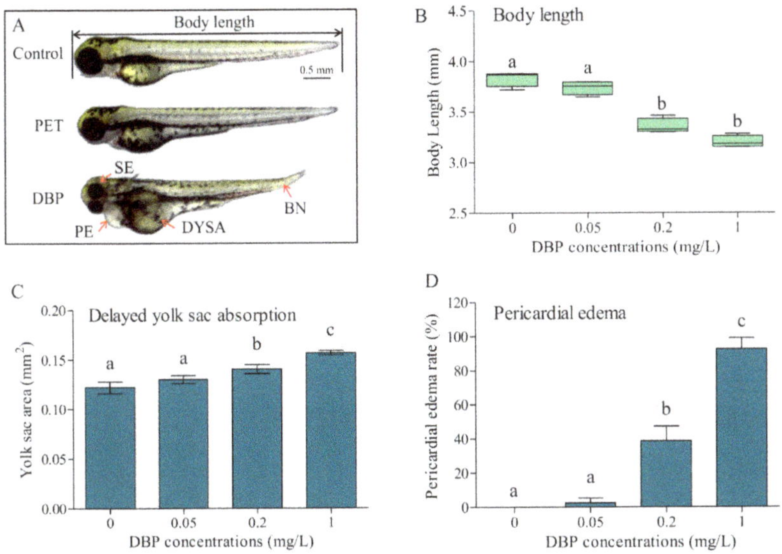

Figure 3. Toxicity effects of PET and DBP singly on the development of zebrafish embryos. (**A**) Embryos under blank control, PET, and DBP expositions; (**B**) Body length of embryos at 72 hpf after DBP exposition; (**C**) Yolk sac area of embryos at 72 hpf after DBP exposition; (**D**) Pericardial edema rates in embryos at 72 hpf after DBP exposition. The letters above the bars indicate significant differences ($p < 0.05$). If two arbitrary groups have the same letter, it signifies that there is no significant difference between them based on the chosen significance level ($p < 0.05$). Abbreviations: BN, bent notochord; DYSA, delayed yolk sac absorption; PE, pericardial edema; SE, small eyes.

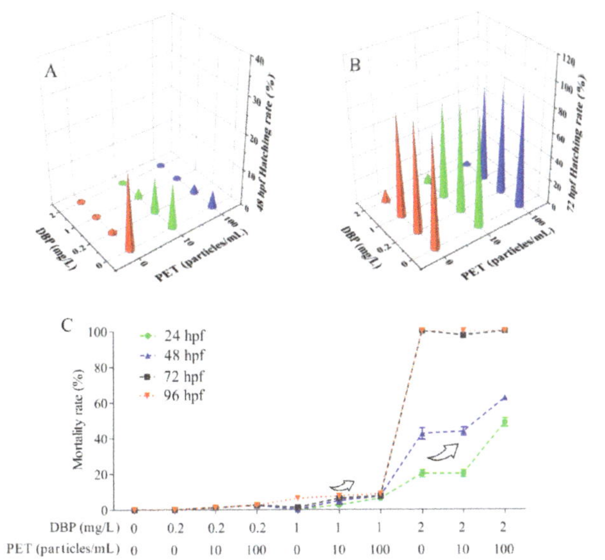

Figure 4. Hatching and mortality rates in zebrafish embryos after co-exposure of DBP + PET. (**A**) Hatching rate at 48 hpf with DBP + PET expositions; (**B**) Hatching rate at 72 hpf with DBP + PET expositions; (**C**) Mortality rate at 24 hpf, 48 hpf, 72 hpf, and 96 hpf with DBP + PET expositions.

After combined exposure to PET, the hatching of zebrafish embryos in 0.2 mg/L DBP treatment was promoted at 48 hpf (Figure 5A). No new malformation phenotype was induced in zebrafish embryos after DBP + PET co-exposure at 48 hpf (Figure 5A). The malformation phenotypes of zebrafish embryos are more diverse and visually apparent at 72 hpf (Figure 5B). The malformation phenotype and delayed yolk sac absorption can still be observed at 72 hpf after DBP exposure. The malformation phenotype, bent notochord, and delayed yolk sac absorption became more severe in 1 mg/L DBP exposition with the co-exposure of 100 particles/mL PET at 72 hpf. The body length of zebrafish embryos in 1 mg/L DBP exposition decreased after co-exposure with 0–100 particles/mL PET at 72 hpf.

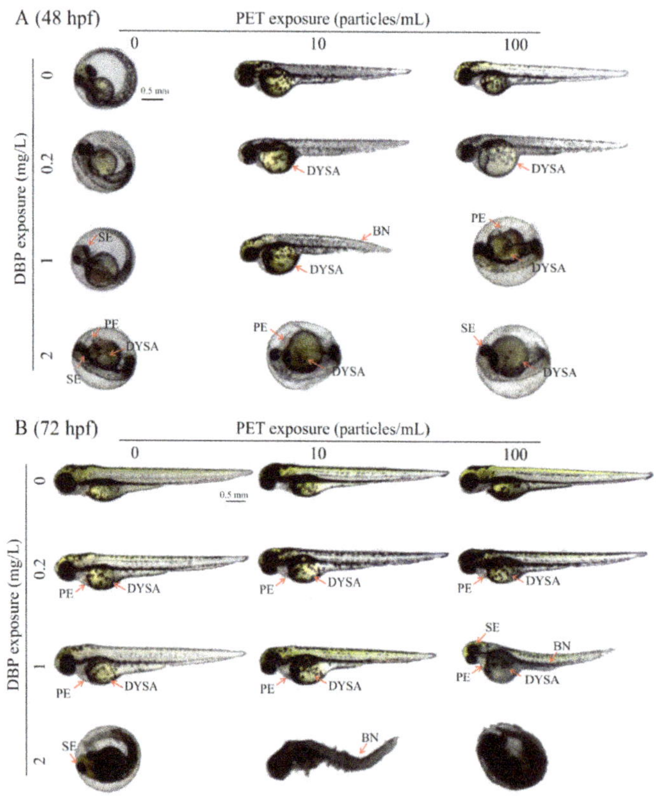

Figure 5. Toxicity effects of DBP + PET co-exposure on the development of zebrafish embryos at 48 hpf (**A**) and 72 hpf (**B**). Abbreviations: BN, bent notochord; DYSA, delayed yolk sac absorption; PE, pericardial edema; SE, small eyes.

4. Discussion

In the present study, single exposure to PET did not significantly affect the growth and survival of zebrafish embryos, but delayed hatching at 48 hpf. Zebrafish embryos were surrounded by chorion until the zebrafish larvae hatched out. The chorion of embryos allows small molecules such as water molecules, ions, and oxygen to enter the cells through the membrane pores and prevents large particles of pollutants from entering the cells [61]. Our results showed that the chorionic surface was partly covered by PET particles (Figure 1C). Duan et al. hypothesized that PET particles may affect the permeability of chorionic channels, thereby reducing oxygen delivery, resulting in an anoxic internal microenvironment in zebrafish embryos [62]. PET adsorption can also enhance the mechanical properties of the embryo's chorionic membrane, which delays embryo hatching [50]. Malafaia et al.

found that zebrafish embryos exposed to polyethylene (PE) microplastics induced early hatching compared to controls [14]. The authors speculated this could be due to chorion damage or changes in water quality, such as induction of hypoxia leading to early hatching, but it was not confirmed; however, premature larvae released into the exposed medium did not survive for long. Furthermore, some studies found that larval fish exposed to PE MPs and Polystyrene (PS) nanoplastics also exhibited malformed phenotypes, such as increased yolk sac area, higher head height, pericardium/yolk sac edema, spine curvature, caudal flexure, and larger optic vesicle area [14,63–65], but PET exposure did not induce malformed phenotypes in zebrafish larvae, which may be due to different toxic mechanisms of different types of MPs. These adverse effects on early life stages of zebrafish suggest that microplastic pollution poses risks to the development and survival of fish and other aquatic organisms.

Exposure to DBP significantly inhibited embryo hatching (≥1 mg/L) and resulted in severe lethal (≥1 mg/L) and teratogenic effects (≥0.2 mg/L). Sun and Li et al. suggested the inhibition of embryo hatching may be due to DBP inhibiting the secretory function of HCGs, leading to the decrease of hatching enzyme secretion and activity [33]. The prominent phenotypes induced by DBP exposure included delayed yolk sac absorption and pericardial edema. The results of body length showed that DBP exposure had significant inhibitory effects on zebrafish embryo growth. Yolk sac is an important source of nutrition for embryo or larva; physical size will gradually decrease with embryonic development [66]. Delayed yolk sac absorption may result in insufficient nutrient supply, which will inhibit the normal growth of zebrafish. Sun and Li showed that exposure to DBP (1.8, 3.6 μM) led to significantly lower heart rate in zebrafish embryos, which was probably related to the malformations pericardial edema and cardiac structure deformities [33]. It is found that 0.6 mg/L BBP significantly increased the malformation rate, caused growth inhibition, increased the cardiac malformation rate, and reduced the heart rate of embryos [67]. The yolk sac and heart may be the priority targets for DBP toxicity.

Mortality increased in co-treatment with 100 particles/mL PET and 2 mg/L DBP at 24 hpf and 48 hpf. The malformation phenotype, bent notochord, and delayed yolk sac absorption became more severe in 1 mg/L DBP exposition with the co-exposure of 100 particles/mL PET at 72 hpf. As previously mentioned, PET has high hydrophobic properties and can easily accumulate on the surface of embryonic chorion. DBP can adsorb to PET through van der Waals forces and hydrophobic interactions [68,69]. When ingested by marine animals, the DBP adsorbed on PET microplastics may desorb into their tissues, causing endocrine disruption and other adverse effects [26,70]. Cao et al. investigated release behaviors of PAEs from twelve microplastics and found that DBP had the strongest release ability in PA, PP, and PET microplastics (47–84%), and they predicted that about approximately 57.8–16,100 kg/year of PAEs are released into the oceans from microplastics [71]. Thus, DBP might be carried to the embryonic chorion by PET. As PET plastics continue to accumulate in the ocean, they may transport increasing amounts of DBP and exacerbate its ecological impacts. DBP is highly lipophilic and can easily pass through the chorion, resulting in high DBP content in embryos.

The interaction between DBP and PET after being taken up by zebrafish needs to be further investigated. Zhang et al. reported that MPs increased the developmental toxicity of cadmium (Cd) on zebrafish embryos but reduced the lethal toxicity of Cd [3]. The surface characteristics and morphology of PET MPs, such as fibers vs. particles, may influence their affinity for and effects on the embryonic chorionic membrane differently. This could result in different toxic effects on the development of waterborne embryos [50]. The findings from Zhang et al. suggest that interactions between chemicals and MPs are complex, and the toxicity to aquatic organisms depends on multiple factors related to the pollutants and physical properties of MPs. In the present study, the 50–200 particles/mL PET MPs delayed hatching and enhanced 1 mg/L DBP-induced bent notochord and delayed yolk sac absorption phenotypes at 72 hpf, but did not directly cause mortality in zebrafish embryos. PET fibers and particles can differ in specific surface area, which may lead to different levels of adsorption and bioconcentration of DBP. The larger surface area of PET particles could facilitate greater adsorption of DBP compared to fibers, making DBP more available to permeate the chorion upon co-exposure. The toxicity of combined exposure to DBP and PET MPs may thus depend on the type of PET (fiber vs. particle), concentration, and duration of exposure. Further research is required to systematically determine how these parameters influence the joint toxicity of MPs and organic pollutants to aquatic life.

5. Conclusions

Our study revealed that PET particles partially covered the embryonic chorion, leading to a decreased hatching rate at 48 hpf. However, PET exposure did not have a significant impact on the growth and survival of zebrafish embryos. In contrast, exposure to DBP caused severe lethal and teratogenic effects, with delayed yolk sac absorption and pericardial edema being the prominent phenotypes. These findings suggest that DBP toxicity primarily targets these two factors. Co-treatment with 100 particles/mL PET and 2 mg/L DBP increased mortality rates at 24 hpf and 48 hpf and resulted in more severe malformation phenotypes such as bent notochords and delayed yolk sac absorption at 72 hpf when exposed to 1 mg/L DBP with the co-exposure of 100 particles/mL PET. It is possible that DBP was carried to the embryonic chorion by PET, which may have increased its bioavailability and, therefore, its toxic effects.

Author Contributions: Methodology, Q.Z., W.M. and J.Z.; software, Q.Z. and J.Z.; validation, Q.Z. and J.Z.; formal analysis, J.Z.; investigation, Q.Z., W.M. and J.Z.; resources, J.Z.; data curation, Q.Z. and W.M.; writing—original draft preparation, Q.Z.; writing—review and editing, J.Z.; visualization, J.Z. and Q.Z.; supervision, J.Z.; project administration, J.Z. All authors have read and agreed to the published version of the manuscript.

Funding: This research was funded by the Natural Science Foundation of China (42106149), the Guangxi Key Research and Development Program of China (GUIKE AB20297018), Guangxi Education Department of China (2021KY0441), and the Science and Technology Base and Talent Project of Guangxi Province (GUIKE AD19245025).

Institutional Review Board Statement: Not applicable.

Informed Consent Statement: Not applicable.

Data Availability Statement: Not applicable.

Conflicts of Interest: The authors declare no conflict of interest.

References

1. Gong, J.; Xie, P. Research progress in sources, analytical methods, eco-environmental effects, and control measures of microplastics. *Chemosphere* **2020**, *254*, 126790. [CrossRef]
2. Eriksen, M.; Lebreton, L.; Carson, H.S.; Thiel, M.; Moore, C.J.; Borerro, J.C.; Galgani, F.; Ryan, P.G.; Reisser, J. Plastic Pollution in the World's Oceans: More than 5 Trillion Plastic Pieces Weighing over 250,000 Tons Afloat at Sea. *PLoS ONE* **2014**, *9*, e111913. [CrossRef] [PubMed]
3. Zhang, R.; Wang, M.; Chen, X.P.; Yang, C.M.; Wu, L.L. Combined toxicity of microplastics and cadmium on the zebrafish embryos (Danio rerio). *Sci. Total Environ.* **2020**, *743*, 140638. [CrossRef]
4. He, Y.; Fan, G.J.; Wu, C.E.; Kou, X.H.; Li, T.T.; Tian, F.; Gong, H. Influence of packaging materials on postharvest physiology and texture of garlic cloves during refrigeration storage. *Food Chem.* **2019**, *298*, 125019. [CrossRef] [PubMed]
5. Jemec, A.; Horvat, P.; Kunej, U.; Bele, M.; Krzan, A. Uptake and effects of microplastic textile fibers on freshwater crustacean Daphnia magna. *Environ. Pollut.* **2016**, *219*, 201–209. [CrossRef] [PubMed]
6. Atugoda, T.; Vithanage, M.; Wijesekara, H.; Bolan, N.; Sarmah, A.K.; Bank, M.S.; You, S.M.; Ok, Y.S. Interactions between microplastics, pharmaceuticals and personal care products: Implications for vector transport. *Environ. Int.* **2021**, *149*, 106367. [CrossRef]
7. Holmes, L.A.; Turner, A.; Thompson, R.C. Interactions between trace metals and plastic production pellets under estuarine conditions. *Mar. Chem.* **2014**, *167*, 25–32. [CrossRef]
8. Holmes, L.A.; Turner, A.; Thompson, R.C. Adsorption of trace metals to plastic resin pellets in the marine environment. *Environ. Pollut.* **2012**, *160*, 42–48. [CrossRef]

9. Fonte, E.; Ferreira, P.; Guilhermino, L. Temperature rise and microplastics interact with the toxicity of the antibiotic cefalexin to juveniles of the common goby (*Pomatoschistus microps*): Post-exposure predatory behaviour, acetylcholinesterase activity and lipid peroxidation. *Aquat. Toxicol.* **2016**, *180*, 173–185. [CrossRef]
10. Ashton, K.; Holmes, L.; Turner, A. Association of metals with plastic production pellets in the marine environment. *Mar. Pollut. Bull.* **2010**, *60*, 2050–2055. [CrossRef]
11. Mao, Y.; Ai, H.; Chen, Y.; Zhang, Z.; Zeng, P.; Kang, L.; Li, W.; Gu, W.; He, Q.; Li, H. Phytoplankton response to polystyrene microplastics: Perspective from an entire growth period. *Chemosphere* **2018**, *208*, 59–68. [CrossRef] [PubMed]
12. Wu, X.; Pan, J.; Li, M.; Li, Y.; Bartlam, M.; Wang, Y. Selective enrichment of bacterial pathogens by microplastic biofilm. *Water Res.* **2019**, *165*, 114979. [CrossRef] [PubMed]
13. Lu, K.; Qiao, R.; An, H.; Zhang, Y. Influence of microplastics on the accumulation and chronic toxic effects of cadmium in zebrafish (*Danio rerio*). *Chemosphere* **2018**, *202*, 514–520. [CrossRef] [PubMed]
14. Malafaia, G.; de Souza, A.M.; Pereira, A.C.; Gonçalves, S.; Da Costa Araújo, A.P.; Ribeiro, R.X.; Rocha, T.L. Developmental toxicity in zebrafish exposed to polyethylene microplastics under static and semi-static aquatic systems. *Sci. Total Environ.* **2020**, *700*, 134867. [CrossRef]
15. Qiang, L.; Cheng, J. Exposure to microplastics decreases swimming competence in larval zebrafish (Danio rerio). *Ecotox. Environ. Saf.* **2019**, *176*, 226–233. [CrossRef]
16. Qiao, R.; Sheng, C.; Lu, Y.; Zhang, Y.; Ren, H.; Lemos, B. Microplastics induce intestinal inflammation, oxidative stress, and disorders of metabolome and microbiome in zebrafish. *Sci. Total Environ.* **2019**, *662*, 246–253. [CrossRef]
17. Jin, Y.; Xia, J.; Pan, Z.; Yang, J.; Wang, W.; Fu, Z. Polystyrene microplastics induce microbiota dysbiosis and inflammation in the gut of adult zebrafish. *Environ. Pollut.* **2018**, *235*, 322–329. [CrossRef]
18. Wan, Z.; Wang, C.; Zhou, J.; Shen, M.; Wang, X.; Fu, Z.; Jin, Y. Effects of polystyrene microplastics on the composition of the microbiome and metabolism in larval zebrafish. *Chemosphere* **2019**, *217*, 646–658. [CrossRef]
19. Chen, Q.; Gundlach, M.; Yang, S.; Jiang, J.; Velki, M.; Yin, D.; Hollert, H. Quantitative investigation of the mechanisms of microplastics and nanoplastics toward zebrafish larvae locomotor activity. *Sci. Total Environ.* **2017**, *584–585*, 1022–1031. [CrossRef]
20. Qiao, R.; Lu, K.; Deng, Y.; Ren, H.; Zhang, Y. Combined effects of polystyrene microplastics and natural organic matter on the accumulation and toxicity of copper in zebrafish. *Sci. Total Environ.* **2019**, *682*, 128–137. [CrossRef]
21. Yu, J.; Chen, L.; Gu, W.; Liu, S.; Wu, B. Heterogeneity effects of nanoplastics and lead on zebrafish intestinal cells identified by single-cell sequencing. *Chemosphere* **2022**, *289*, 133133. [CrossRef] [PubMed]
22. Mak, C.W.; Ching-Fong Yeung, K.; Chan, K.M. Acute toxic effects of polyethylene microplastic on adult zebrafish. *Ecotox. Environ. Saf.* **2019**, *182*, 109442. [CrossRef] [PubMed]
23. Lemoine, C.M.R.; Kelleher, B.M.; Lagarde, R.; Northam, C.; Elebute, O.O.; Cassone, B.J. Transcriptional effects of polyethylene microplastics ingestion in developing zebrafish (Danio rerio). *Environ. Pollut.* **2018**, *243*, 591–600. [CrossRef] [PubMed]
24. Wang, P.; Zhao, Y. Plasticizer Exposure and Reproductive Health: Phthalates and Bisphenol A. In *Emerging Chemicals and Human Health*; Zhang, Y., Ed.; Springer: Singapore, 2019; pp. 49–67. ISBN 978-981-32-9535-3.
25. Cormier, B.; Cachot, J.; Blanc, M.; Cabar, M.; Clérandeau, C.; Dubocq, F.; Le Bihanic, F.; Morin, B.; Zapata, S.; Begout, M.; et al. Environmental microplastics disrupt swimming activity in acute exposure in Danio rerio larvae and reduce growth and reproduction success in chronic exposure in D. rerio and Oryzias melastigma. *Environ. Pollut.* **2022**, *308*, 119721. [CrossRef]
26. Rochman, C.M.; Kurobe, T.; Flores, I.; Teh, S.J. Early warning signs of endocrine disruption in adult fish from the ingestion of polyethylene with and without sorbed chemical pollutants from the marine environment. *Sci. Total Environ.* **2014**, *493*, 656–661. [CrossRef]
27. Burton, G.A. Stressor Exposures Determine Risk: So, Why Do Fellow Scientists Continue To Focus on Superficial Microplastics Risk? *Environ. Sci. Technol.* **2017**, *51*, 13515–13516. [CrossRef]
28. Koelmans, A.A.; Besseling, E.; Foekema, E.; Kooi, M.; Mintenig, S.; Ossendorp, B.C.; Redondo-Hasselerharm, P.E.; Verschoor, A.; van Wezel, A.P.; Scheffer, M. Risks of Plastic Debris: Unravelling Fact, Opinion, Perception, and Belief. *Environ. Sci. Technol.* **2017**, *51*, 11513–11519. [CrossRef]
29. Heudorf, U.; Mersch-Sundermann, V.; Angerer, J. Phthalates: Toxicology and exposure. *Int. J. Hyg. Envir. Health* **2007**, *210*, 623–634. [CrossRef]
30. Schettler, T. Human exposure to phthalates via consumer products. *Int. J. Androl.* **2006**, *29*, 134–139. [CrossRef]
31. Hauser, R.; Duty, S.; Godfrey-Bailey, L.; Calafat, A.M. Medications as a source of human exposure to phthalates. *Environ. Health Perspect.* **2004**, *112*, 751–753. [CrossRef]
32. Schecter, A.; Lorber, M.; Guo, Y.; Wu, Q.; Yun, S.H.; Kannan, K.; Hommel, M.; Imran, N.; Hynan, L.S.; Cheng, D.; et al. Phthalate Concentrations and Dietary Exposure from Food Purchased in New York State. *Environ. Health Perspect.* **2013**, *121*, 473–479. [CrossRef] [PubMed]
33. Sun, G.; Li, Y. Exposure to DBP induces the toxicity in early development and adverse effects on cardiac development in zebrafish (*Danio rerio*). *Chemosphere* **2019**, *218*, 76–82. [CrossRef] [PubMed]
34. Bang, D.Y.; Lee, I.K.; Lee, B. Toxicological Characterization of Phthalic Acid. *Toxicol. Res.* **2011**, *27*, 191–203. [CrossRef] [PubMed]
35. Chatterjee, S.; Karlovsky, P. Removal of the endocrine disrupter butyl benzyl phthalate from the environment. *Appl. Microbiol. Biot.* **2010**, *87*, 61–73. [CrossRef]

36. Cotruvo, J.A. 2017 WHO Guidelines for Drinking Water Quality: First Addendum to the Fourth Edition. *J. AWWA* **2017**, *109*, 44–51. [CrossRef]
37. Di Lorenzo, M.; Forte, M.; Valiante, S.; Laforgia, V.; De Falco, M. Interference of dibutylphthalate on human prostate cell viability. *Ecotox. Environ. Saf.* **2018**, *147*, 565–573. [CrossRef] [PubMed]
38. Staples, C.A.; Peterson, D.R.; Parkerton, T.F.; Adams, W.J. The environmental fate of phthalate esters: A literature review. *Chemosphere* **1997**, *35*, 667–749. [CrossRef]
39. Högberg, J.; Hanberg, A.; Berglund, M.; Skerfving, S.; Remberger, M.; Calafat, A.M.; Filipsson, A.F.; Jansson, B.; Johansson, N.; Appelgren, M.; et al. Phthalate Diesters and Their Metabolites in Human Breast Milk, Blood or Serum, and Urine as Biomarkers of Exposure in Vulnerable Populations. *Environ. Health Perspect.* **2008**, *116*, 334–339. [CrossRef]
40. Oehlmann, J.; Oetken, M.; Schulte-Oehlmann, U. A critical evaluation of the environmental risk assessment for plasticizers in the freshwater environment in Europe, with special emphasis on bisphenol A and endocrine disruption. *Environ. Res.* **2008**, *108*, 140–149. [CrossRef]
41. Liu, H.; Cui, K.; Zeng, F.; Chen, L.; Cheng, Y.; Li, H.; Li, S.; Zhou, X.; Zhu, F.; Ouyang, G.; et al. Occurrence and distribution of phthalate esters in riverine sediments from the Pearl River Delta region, South China. *Mar. Pollut. Bull.* **2014**, *83*, 358–365. [CrossRef]
42. Chen, H.; Mao, W.; Shen, Y.; Feng, W.; Mao, G.; Zhao, T.; Yang, L.; Yang, L.; Meng, C.; Li, Y.; et al. Distribution, source, and environmental risk assessment of phthalate esters (PAEs) in water, suspended particulate matter, and sediment of a typical Yangtze River Delta City, China. *Environ. Sci. Pollut. Res.* **2019**, *26*, 24609–24619. [CrossRef] [PubMed]
43. Zhu, Q.; Xu, L.; Wang, W.; Liu, W.; Liao, C.; Jiang, G. Occurrence, spatial distribution and ecological risk assessment of phthalate esters in water, soil and sediment from Yangtze River Delta, China. *Sci. Total Environ.* **2022**, *806*, 150966. [CrossRef] [PubMed]
44. Lee, Y.; Lee, J.; Choe, W.; Kim, T.; Lee, J.; Kho, Y.; Choi, K.; Zoh, K. Distribution of phthalate esters in air, water, sediments, and fish in the Asan Lake of Korea. *Environ. Int.* **2019**, *126*, 635–643. [CrossRef] [PubMed]
45. Pu, S.; Hamid, N.; Ren, Y.; Pei, D. Effects of phthalate acid esters on zebrafish larvae: Development and skeletal morphogenesis. *Chemosphere* **2020**, *246*, 125808. [CrossRef]
46. Qian, L.; Liu, J.; Lin, Z.; Chen, X.; Yuan, L.; Shen, G.; Yang, W.; Wang, D.; Huang, Y.; Pang, S.; et al. Evaluation of the spinal effects of phthalates in a zebrafish embryo assay. *Chemosphere* **2020**, *249*, 126144. [CrossRef]
47. Jergensen, T.; Cusmano, D.; Roy, N.M. Di-butyl phthalate (DBP) induces craniofacial defects during embryonic development in zebrafish. *Ecotoxicology* **2019**, *28*, 995–1002. [CrossRef]
48. Mu, X.; Huang, Y.; Li, J.; Yang, K.; Yang, W.; Shen, G.; Li, X.; Lei, Y.; Pang, S.; Wang, C.; et al. New insights into the mechanism of phthalate-induced developmental effects. *Environ. Pollut.* **2018**, *241*, 674–683. [CrossRef]
49. Barbagallo, S.; Baldauf, C.; Orosco, E.; Roy, N.M. Di-butyl phthalate (DBP) induces defects during embryonic eye development in zebrafish. *Ecotoxicology* **2022**, *31*, 178–185. [CrossRef]
50. Cheng, H.; Feng, Y.; Duan, Z.; Duan, X.; Zhao, S.; Wang, Y.; Gong, Z.; Wang, L. Toxicities of microplastic fibers and granules on the development of zebrafish embryos and their combined effects with cadmium. *Chemosphere* **2021**, *269*, 128677. [CrossRef]
51. Godoy, V.; Blázquez, G.; Calero, M.; Quesada, L.; Martín-Lara, M.A. The potential of microplastics as carriers of metals. *Environ. Pollut.* **2019**, *255*, 113363. [CrossRef]
52. Rivera-Hernández, J.R.; Fernández, B.; Santos-Echeandia, J.; Garrido, S.; Morante, M.; Santos, P.; Albentosa, M. Biodynamics of mercury in mussel tissues as a function of exposure pathway: Natural vs microplastic routes. *Sci. Total Environ.* **2019**, *674*, 412–423. [CrossRef] [PubMed]
53. Chen, Q.; Yin, D.; Jia, Y.; Schiwy, S.; Legradi, J.; Yang, S.; Hollert, H. Enhanced uptake of BPA in the presence of nanoplastics can lead to neurotoxic effects in adult zebrafish. *Sci. Total Environ.* **2017**, *609*, 1312–1321. [CrossRef] [PubMed]
54. Qi, Z.; Chen, M.; Song, Y.; Wang, X.; Li, B.; Chen, Z.; Tsang, S.Y.; Cai, Z. Acute exposure to triphenyl phosphate inhibits the proliferation and cardiac differentiation of mouse embryonic stem cells and zebrafish embryos. *J. Cell. Physiol.* **2019**, *234*, 21235–21248. [CrossRef] [PubMed]
55. Belanger, S.E.; Rawlings, J.M.; Carr, G.J. Use of fish embryo toxicity tests for the prediction of acute fish toxicity to chemicals. *Environ. Toxicol. Chem.* **2013**, *32*, 1768–1783. [CrossRef]
56. Lammer, E.; Carr, G.J.; Wendler, K.; Rawlings, J.M.; Belanger, S.E.; Braunbeck, T. Is the fish embryo toxicity test (FET) with the zebrafish (*Danio rerio*) a potential alternative for the fish acute toxicity test? *Comp. Biochem. Physiol. Part C Toxicol. Pharmacol.* **2009**, *149*, 196–209. [CrossRef]
57. Aldavood, S.J.; Abbott, L.C.; Evans, Z.R.; Griffin, D.J.; Lee, M.D.; Quintero-Arevalo, N.M.; Villalobos, A.R. Effect of Cadmium and Nickel Exposure on Early Development in Zebrafish (*Danio rerio*) Embryos. *Water-Sui.* **2020**, *12*, 3005. [CrossRef]
58. Buschmann, J. The OECD Guidelines for the Testing of Chemicals and Pesticides. In *Teratogenicity Testing: Methods and Protocols*; Barrow, P.C., Ed.; Humana Press: Totowa, NJ, USA, 2013; pp. 37–56. ISBN 978-1-62703-131-8.
59. Su, L.; Xue, Y.; Li, L.; Yang, D.; Kolandhasamy, P.; Li, D.; Shi, H. Microplastics in Taihu Lake, China. *Environ. Pollut.* **2016**, *216*, 711–719. [CrossRef]
60. Silva Brito, R.; Canedo, A.; Farias, D.; Rocha, T.L. Transgenic zebrafish (Danio rerio) as an emerging model system in ecotoxicology and toxicology: Historical review, recent advances, and trends. *Sci. Total Environ.* **2022**, *848*, 157665. [CrossRef]
61. Mu, X.Y.; Qi, S.Z.; Liu, J.; Yuan, L.L.; Huang, Y.; Xue, J.Y.; Qian, L.; Wang, C.J.; Li, Y.R. Toxicity and behavioral response of zebrafish exposed to combined microplastic and bisphenol analogues. *Environ. Chem. Lett.* **2022**, *20*, 41–48. [CrossRef]

62. Duan, Z.; Duan, X.; Zhao, S.; Wang, X.; Wang, J.; Liu, Y.; Peng, Y.; Gong, Z.; Wang, L. Barrier function of zebrafish embryonic chorions against microplastics and nanoplastics and its impact on embryo development. *J. Hazard. Mater.* **2020**, *395*, 122621. [CrossRef]
63. Pitt, J.A.; Kozal, J.S.; Jayasundara, N.; Massarsky, A.; Trevisan, R.; Geitner, N.; Wiesner, M.; Levin, E.D.; Di Giulio, R.T. Uptake, tissue distribution, and toxicity of polystyrene nanoparticles in developing zebrafish (*Danio rerio*). *Aquat. Toxicol.* **2018**, *194*, 185–194. [CrossRef] [PubMed]
64. Bhagat, J.; Zang, L.; Nishimura, N.; Shimada, Y. Zebrafish: An emerging model to study microplastic and nanoplastic toxicity. *Sci. Total Environ.* **2020**, *728*, 138707. [CrossRef] [PubMed]
65. Sökmen, T.Ö.; Sulukan, E.; Türkoğlu, M.; Baran, A.; Özkaraca, M.; Ceyhun, S.B. Polystyrene nanoplastics (20 nm) are able to bioaccumulate and cause oxidative DNA damages in the brain tissue of zebrafish embryo (*Danio rerio*). *Neurotoxicology* **2020**, *77*, 51–59. [CrossRef] [PubMed]
66. Zhang, Q.; Cheng, J.P.; Xin, Q. Effects of tetracycline on developmental toxicity and molecular responses in zebrafish (Danio rerio) embryos. *Ecotoxicology* **2015**, *24*, 707–719. [CrossRef]
67. Sun, G.; Liu, K. Developmental toxicity and cardiac effects of butyl benzyl phthalate in zebrafish embryos. *Aquat. Toxicol.* **2017**, *192*, 165–170. [CrossRef]
68. Mato, Y.; Isobe, T.; Takada, H.; Kanehiro, H.; Ohtake, C.; Kaminuma, T. Plastic Resin Pellets as a Transport Medium for Toxic Chemicals in the Marine Environment. *Environ. Sci. Technol.* **2001**, *35*, 318–324. [CrossRef]
69. Fu, L.; Li, J.; Wang, G.; Luan, Y.; Dai, W. Adsorption behavior of organic pollutants on microplastics. *Ecotox. Environ. Saf.* **2021**, *217*, 112207. [CrossRef]
70. Teuten, E.L.; Saquing, J.M.; Knappe, D.R.U.; Barlaz, M.A.; Jonsson, S.; Björn, A.; Rowland, S.J.; Thompson, R.C.; Galloway, T.S.; Yamashita, R.; et al. Transport and release of chemicals from plastics to the environment and to wildlife. *Philos. Trans. R. Soc. B Biol. Sci.* **2009**, *364*, 2027–2045. [CrossRef]
71. Cao, Y.; Lin, H.; Zhang, K.; Xu, S.; Yan, M.; Leung, K.M.Y.; Lam, P.K.S. Microplastics: A major source of phthalate esters in aquatic environments. *J. Hazard. Mater.* **2022**, *432*, 128731. [CrossRef]

toxics

MDPI

Review

A Hidden Pathway for Human Exposure to Micro- and Nanoplastics—The Mechanical Fragmentation of Plastic Products during Daily Use

Yang Yu [1], Nicholas Craig [2] and Lei Su [1,3,*]

[1] College of Marine Ecology and Environment, Shanghai Ocean University, Shanghai 201306, China
[2] School of Biosciences, The University of Melbourne, Parkville, VIC 3010, Australia
[3] Shanghai Engineering Research Center of River and Lake Biochain Construction and Resource Utilization, Shanghai 201702, China
* Correspondence: lsu@shou.edu.cn

Abstract: In numerous environmental compartments around the world, the existence of micro- and nanoplastics (MNPs) in the environment has been verified. A growing number of studies have looked at the interaction between MNPs and human activities due to the risks they may pose to humans. Exposure pathways are key factors in measuring MNPs risks. However, current research largely ignores the contribution of mechanical fragmentation pathways to MNPs exposure during the daily use of plastic products. Our critical review demonstrated the research gap between MNP fragmentation and risk assessments via a network analysis. The release of fragmented MNPs and their properties were also described at various scales, with emphasis on environmental stressors and mechanical fragmentation. In the scenarios of daily use, plastic products such as food packaging and clothing provide acute pathways of MNPs exposure. The release tendency of those products (up to 10^2 mg MNPs) are several orders of magnitude higher than MNPs abundances in natural compartments. Despite the limited evidence available, waste recycling, landfill and municipal activities represented long-term pathways for MNPs fragmentation and point sources of MNPs pollution in environmental media. Assessing the health effects of the fragmentation process, unfortunately, is further hampered by the current absence of human exposure impact assessments for secondary MNPs. We proposed that future studies should integrate aging evaluation into risk assessment frameworks and establish early warning signs of MNPs released from plastic products.

Keywords: microplastic; pathways; mechanical fragmentation; exposure

1. Introduction

As humanity enters the so-called plastic age, plastic products have pervaded almost every aspect of human life [1–3]. The toxicological impacts and ecological risks of the aforementioned compounds are being thoroughly assessed, particularly since microplastics and nanoplastics are of growing concern in environmental science and public health [4,5]. They have also been shown to have acute and long-term harmful effects in a range of models, including plankton and mammals, with adverse consequences ranging from individual mortality to behavioral abnormalities [6–8]. Whether micro- and nanoplastics (MNPs) at environmentally relevant conditions can likewise pose major dangers is, of course, still up for debate [9,10].

Humans are also inevitably exposed to MNPs in the environment via different pathways. Due to the early beginnings of plastic debris monitoring in marine organisms, a significant portion of which are considered as animal protein, there is a long history of research on the estimation of human exposure to MNPs via seafood consumption, typically of fish and mussels [11–13]. The risk of exposure via aquatic products is related to the specific part of the aquatic product consumed, e.g., MNPs contamination in the gastrointestinal tract and gills of fish does not fully reflect the risk of exposure [14].

In the field of public health research, the study of MNPs exposure pathways is more diverse, in that, in addition to the oral exposure pathways such as eating food and drinking water, a considerable number of studies have been conducted on atmospheric MNPs concentration levels and respiratory exposure pathways [15,16]. In terms of abundances, MNPs in the atmosphere can reach up to four orders of magnitude higher in indoor and urban environments, far exceeding the levels at which they have been detected in the same volume of aqueous sediments and organisms [17,18]. The great majority of these high levels of MNPs in the environment around humans are caused by the direct release of plastic products. This indicates that a significant amount of human exposure to MNPs may result from the regular usage of plastic products.

The sources of MNPs have usually been reported qualitatively or semi-quantitatively during the examination of exposure pathways in both indoor and field experiments. For instance, whereas bio-fragmentation during ingestion contributes to a minor portion of the MNPs in fish, in the calculations, they are all believed to originate from the environment [19]. Also, weathered plastic is frequently employed in exposure investigations to increase the environmental relevance of the data [20,21]; however, the secondary MNPs produced by these weathered plastics are rarely used for concentration corrections when calibrating concentrations. In fact, the majority of MNPs in the environment are secondary byproducts of the fragmentation of large items, with the exception of a small number of primary MNPs, such as pellets and micro-spheres [22,23]. Due to the limited production and use of manmade nanoplastics, the fragmentation process contributes almost the entire source of nanoplastics in the environment. Applying the theory of morphology, it has been shown that the surface structure of MNPs changes during fragmentation, for example, by increasing the specific surface area and complexity, that promote both the release of residual contaminants from the plastics and the enhancement of the level of adsorption of these secondary particles onto persistent organic pollutants (POPs) in the environment [24–26]. As a result, the secondary particles produced by the fragmentation of plastics have different physical and chemical properties, which raise the risk to the environment.

As the size of the MNPs observed continues to decrease, more and more focus is being placed on investigations of the fragmentation mechanisms, not the least of which is the application of polymer physics to explain them at the microscopic level [20,27]. Meanwhile, in the current body of toxicological research, the fragmentation of plastic items has rarely been considered as a potential exposure pathway for MNPs. Studies on various plastic products, including pacifiers, masks and drink bottles, have actually shown that a significant amount of MNPs is created by mechanical abrasion or by high temperatures and pressures, and that this quantity rises with usage [28–30]. These plastics come into close contact with humans during daily use until they enter the body via inhalation or food consumption. Unfortunately, the fragmentation of plastics that are directly related to human daily life has received little attention in current research, and toxicological studies using animal models have rarely taken into account the effect of fragmentation of aged plastics on concentration increases during exposure, even though this effect is critical in chronic and in situ experiments. The health risks of human exposure to aging plastics under conditions of daily use can only be completely addressed with a thorough understanding of the plastic fragmentation process and its implications for the production of secondary pollutants. To this end, based on an introduction to the toxicity and fragmentation mechanisms of MNPs, the current work will systematically review: (1) research trends related to plastic fragmentation mechanisms; (2) fragmentation scenarios and mechanisms of plastics products that are closely related to human beings; and (3) the significance of the fragmentation process in terms of the release of pollutants and their health risks.

2. The Research Trends and Knowledge Gaps

In systematic reviews pertaining to pollutants, keyword-based bibliometric analyses have frequently been utilized to provide a visual depiction of research trends and hotspots in a particular research topic. We extracted all of the bibliometric information from 431 research papers using the keywords of fragmentation and microplastic. The search was conducted via WOS core selection on 15 June 2023. A network correlation was built based on the top 50 keywords from those papers (Figure 1). The research on fragmentation clearly focuses more on particular polymer types and degradation mechanisms around the main keywords. Instead, phrases associated with toxicity, exposure impacts, etc. and microplastics are grouped together under one category, and the other clusters also include research on marine litter and chemical adsorption. The aforementioned clustering patterns address the issue that only a small number of recent studies on plastic fragmentation have been taken into account when determining the risks of MNPs. The two core keywords, microplastics and fragmentation, are also split in research (Figure 1, direction of the arrow). Notably, the keywords related to chemisorption were independent of microplastics and fragmentation, suggesting that the environmental risk of MNPs releasing contaminants during fragmentation is missing in most of the toxicological studies. A viewpoint proposed that weathered plastic fulfils two of three criteria to impose a planetary boundary threat related to "chemical pollution and the release of novel entities" [31]. Similar ideas are echoed in some of the early studies of marine litter, that the aging process of plastics is an important contributor to the exacerbation of their ecological risks [23,32,33]. The current study trends make it evident that, in comparison with the aging mechanism, the risk of MNPs exposure derived from the fragmentation process is still not getting the attention it deserves. This could grossly underestimate the ecological risks of plastic debris in the field.

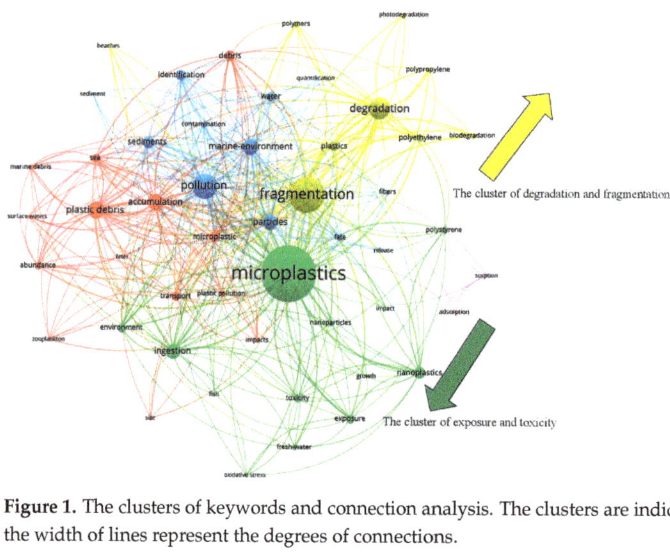

Figure 1. The clusters of keywords and connection analysis. The clusters are indicated by colors and the width of lines represent the degrees of connections.

It is evident from the recent literature reviews that the conditions for fragmentation research are changing from ideal indoor settings to environmentally relevant scenarios. For example, Born et al. [34] concluded that, although fewer than half of all studies were conducted under environmentally relevant conditions, weathering and fragmentation experiments followed the transition from model to nature. Natural and daily-use conditions for plastic fragmentation are increasingly being incorporated into indoor accelerated or in situ aging simulations, and some studies have even gone so far as to directly use plastic fragments aged in the environment for the preparation of MNPs for exposure experiments [35–37]. Due to the lack of environmentally relevant conditions, however, a toxicity baseline of MNPs generated from fragmentation remains unclear. To this end, a full understanding of the roles of fragmentation is a fundamental task.

3. Stress-Induced Plastic Item Fragmentation Leading to the Production of MNPs

3.1. Environmental Stress

In fracture research, environmental stresses are frequently referred to as external forces that eventually induce material failure owing to corrosion brought on by solutions or other media [38]. From the perspective of MNPs, the major environmental stresses that can trigger plastic fragmentation are UV radiation, salt-water infiltration and temperature changes related to meteorological factors [39,40]. They are characterized by a long duration of effect, uniformity of action and diffuse distribution over the surface of the material. Historically, in order to test the performance and weathering resistance of plastic products, the roles of environmental stresses had been key external factors to be considered in the simulation of exposure. A number of in situ aging experiments have been set up directly in different latitudinal regions of the earth to explore the aging properties of plastics under long-term sustained environmental exposure [41]. In contrast to intuition, some thermoplastics can exhibit defects visible to the naked eye, such as cracks and surface delamination, after a relatively short exposure time (<1 month) [42]. These defects demonstrate the possibility of short-term environmental stress-induced formation of MNPs on the surfaces of plastic products.

At various observational scales, polymers subjected to environmental stressors exhibit gradual changes (Figure 2). Within several years of UV exposure, polypropylene, for instance, will initially experience a decrease in the degree of polymerization or a reduction in chain length and molecular weight at the macromolecular level [27]. This modification will cause the polymer's crystalline zone to grow, increasing crystallinity [43,44]. Additionally, it means that the materials would become more fragile and prone to breaking. Since the above-described aging process is irreversible under natural circumstances, different polymer materials' propensity to release MNPs into the environment can potentially be anticipated using their thermodynamic characteristics. For instance, it is known via molecular simulations, e.g., crystallinity and hydrophobicity predictions, that polypropylene ages in artificial seawater more quickly than nylon, which means that polypropylene fishing gear will release MNPs earlier during its use [45]. Again, the formation of environmentally persistent free radicals on plastic surfaces was highly suspected as a trigger for degradation and physical damage [46]. It should be noted that projections based on the polymer properties of materials do not accurately reflect the true lifetime of plastics under complicated conditions of usage, much less the risks of exposure to microplastic emissions from the processes outlined above, due to the complexity of environmental pressures. As the most fundamental factor that causes plastic aging, the contribution of environmental stressors to the release of MNPs under actual-use conditions warrants further research.

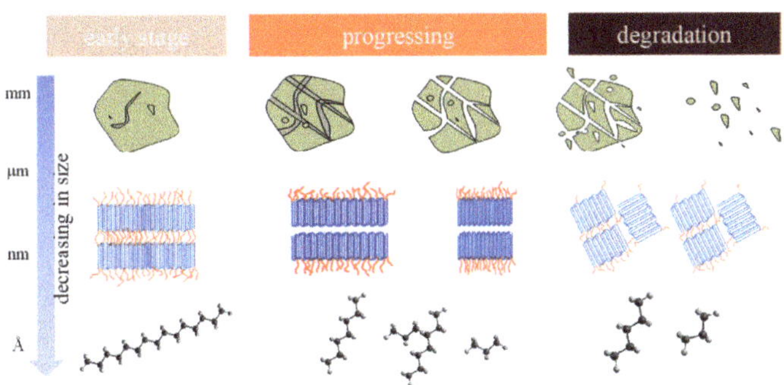

Figure 2. Observing the aging progress of polymers from the molecular (characterization using high-resolution SEM/TEM) to macro-scale (naked eye).

3.2. Mechanical Abrasion and Wear

Mechanical abrasion and wearing damage plastics directly by actions including shearing, compression and tearing, in contrast to environmental stressors. In situ experiments have mostly concentrated on mechanical abrasion under the conditions of natural beach ecosystems, including the contributions of wave wash and sediment corrosion to fragmentation and MNPs generation [47–49]. The above studies have demonstrated that natural abrasion processes in the beach environment are a more important contributor to the nearshore accumulation of MNPs than the discharge of primary MNPs. It is clear that the mechanical abrasion resulting from the daily use of plastics is more intense and immediate than in natural conditions, but related studies in the area of MNPs have not been widely available until only very recently [34]. When we expand the scope of the study to the entire field of public health, the risk of inhalation of dust or particles represented by asbestos and synthetic leather has been systematically studied as early as the 1960s [50,51]. The acute causes of dust inhalation are essentially related to anthropogenic fragmentation, such as construction, metalworking and scrap recycling. They have shown mechanical fragmentation to be a significant contributor to the sources of hazardous compounds, despite the lack of a clear terminology.

The generation of MNPs via mechanical fragmentation is primarily derived from abrasive and adhesive wear, which are two of the most fundamental mechanisms involved in surface abrasion and wearing (Figure 3). For abrasive wearing, MNPs are formed when foreign particles collide with the surface of a plastic, and the direction in which these particles move has an impact on the shape of the MNPs. For instance, the vertical force of the particles can cause the plastic's surface to crack, and, as the crack spreads, porcelain-tile-like secondary fragments are created. This could explain the cross-sectional microstructure and crack patterns of weathered plastic debris [52,53]. The particles tend to cleave into flaky secondary MNPs when they move at an angle or even parallel to the plastic's surface. Such a phenomenon is obvious when inspecting the surfaces of micro-fibers collected from trail-running events [54]. Adhesive wear, in contrast, involves the contact and interaction of asperities on two surfaces with a strong adhesive force. Tire wear is thought to be one of the primary types of MNPs in road dust deposition due to adhesive wear such as braking [55–57]. In daily life, the opening and closing of plastic bottles represent typical adhesive wear and can form fatigue cracks after 10 min with twisting caps [58,59]. Plastic products that have been used or even aged are more susceptible to mechanical fragmentation than brand-new plastic products. Weathered plastic fragments, for instance, can produce more secondary MNPs during the shaking process than newly purchased ones [44,60]. This means that long-term environmental stressors operate as the basis for the majority of pathways for the generation of MNPs from mechanical fragmentation.

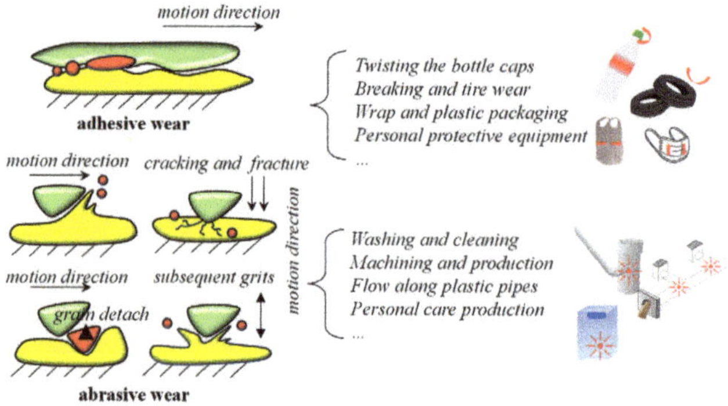

Figure 3. The generation of secondary MNPs in the processes of adhesive and abrasive wear and their representative activities.

4. The Characteristics of Secondary MNPs Released by Fragmentation

4.1. Geometrical Features

The geometrical features of MNPs largely varied by size category (Figure 4). For large particles, they were more likely to be generated via direct fragmentation. This resulted in irregular shapes being created such as sharp, asymmetric and coarse perforations. The structure of crystallization also played a role in shape formation. For example, weathered polyethylene tended to develop more strips than polypropylene due to its lamellar structure [44]. With decreasing size, the fragmented particle shape could be impacted from particle–particle and particle–wall collisions that are similar to fine sediment dynamics [61,62]. That would promote the generation of regular shapes with high degrees of roundness and sphericity. Recent field observations generally agreed with such deductions. In comparison with fiber, more fragments and spheres were found as part of the small microplastics and nanoplastics fraction [63,64]. The ultra-microstructure (1–10 nm) of nanoplastics is equivalent to the chain structures of given polymers to some extent. In this size category, there are substantial agglomerations of polymer chains that display a range of complex structures, e.g., lamellar and spherulitic. This is similar to the case of magnetite nano-crystals [65]. However, the geometrical features of ultra-small nanoplastics remain unclear due to the limited number of observations. The geometrical characteristics of MNPs are frequently associated with bodily harm. For instance, when considering broken plastic products, sharp cracks can scratch the skin and lead to infections. On a smaller scale, microfractures in MNPs may also damage the digestive tracts of organisms via respiratory and feeding pathways [7]. According to reports, smaller microplastics can penetrate cell membranes and cause an inflammatory reaction, which is a much more serious physical injury than that caused by larger plastics [66,67]. Unfortunately, little research has been conducted on the physical harm that MNPs cause to the human body as they fragment.

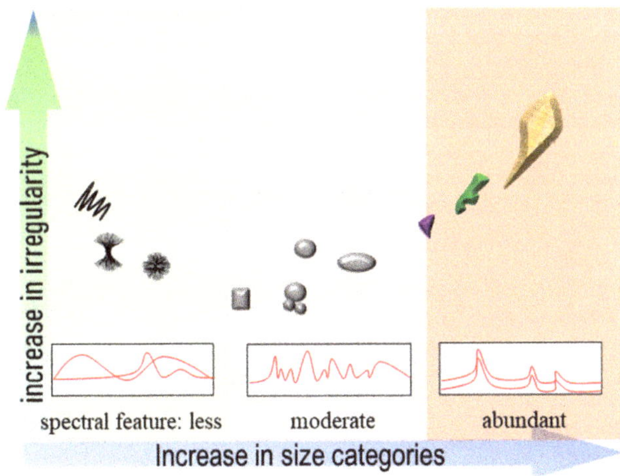

Figure 4. A diagram indicating how MNPs change when we "look" at them at different scales.

4.2. Surface Textures

Surface textures were too often ignored in MNPs characterization although they were closely related to the particle structure and chemical adsorption. If MNPs were directly generated by fractures and the breakdown of base materials, the surface textures denoted the variance of stress load and could be diagnosed using fractography. At this point, the surface textures exhibited certain patterns according to the polymer compositions and local modulus. Regarding nanoplastics in the field, they were primarily generated by embrittled and highly weathered surfaces and could show gradients of embrittlement deformation, e.g., brittle fractures and beach marks [27,53]. However, a smooth and flat surface was expected if particle erosion and collision continued in the weathering and aging processes. This mechanism was quite similar to the evolution of particle geometrical features as the surface patterns tended to be more regular. On the other hand, the surface texture roughly indicated the presence of insoluble material such as inorganic stuffing and additives. That meant that the surfaces of nanoplastics could be observed with embedded foreign matter and could be more distinct for small nanoplastics. Again, the ultra-microstructure of the surface texture is determined using chain structure crystallization. Further improvements in observation technologies are required. The surface texture of MNPs and the absorption of chemical contaminants are intimately connected. The vector effect of MNPs describes how a rougher surface has a higher specific surface area that allows the particles to adsorb more contaminants and deliver them to the body via different pathways [68,69]. The likelihood that POPs may enter the human body via MNPs can be readily identified if the surface textures of the individual particles are accurately quantified. On the other hand, MNPs with higher adsorption capacities may also be able to transport persistent pollutants from the organism's gastrointestinal location to the outside of the body during digestion [68]. In order to prevent mistakes in the assessment of the risk of contaminant release, full partitioning and dissolved equilibrium concentrations should be taken into account when determining the actual risk of MNPs.

5. Human Exposure to MNPs via Mechanical Fragmentation

5.1. The Pathways of Acute Exposure

The acute exposure pathways included the use of all kinds of everyday household items, such as cloth, personal protective equipment, food packaging and personal care products (Table 1). Due to its close relationship to human health and intensive use, more than 50% of the studies focused on food packaging such as teabags and water bottles (Figure 5a–c). This also results in a high level of environmental stress from liquid and heat (Figure 5d). Despite the difference in quantification, the levels of release tendency of those products (up to 10^2 mg MNPs) are several orders of magnitude higher than MNP abundances in natural compartments, e.g., soil, water and biota [5,70,71]. The fragmentation process is more pronounced in the cases of beverage bottles and teabags because of their direct contact with liquids during usage and the potential for heating, which is primarily related to environmental pressures. In a study using controllable conditions, the use of hot water induced 10% more MNPs compared with the use of cold water [72]. Interestingly, not all environmental stressors significantly promote fragmentation. For example, the release of MNPs from rLDPE/LLDPE-modified asphalt during abrasion is quite stable at different water pH levels compared to other environmental factors [73]. Therefore, we must perform a complete screening for the major environmental stressors that contribute to fragmentation in the future. If we set 1000 nm as the benchmark for sub-microplastics and 100 nm for nanoplastics, only a limited number of studies have observed them in all MNPs (Table 1). This could be attributed to the low resolution of the detection techniques employed rather than the absence of smaller particles. Based on the surface fracture characteristics observed using SEM, friction marks can be produced on the surfaces of the particles even by means of washing, agitation, etc., suggesting the release of smaller-sized MNPs [30].

Table 1. The acute exposure of humans to MNPs via multiple production means.

Production	Primary Mechanical Abrasion (Force Analysis)	Secondary MNPs	Release Tendency	Reference
mask	adhesive wear (compression and twisting)	fiber (0.5 mm to 3.8 mm)	24,300–55,900 MNPs/mask/d	[28]
plain woven	adhesive wear (compression and twisting)	fiber (0.1 mm to 1.5 mm)	51.6–107.7 mg MNPs/kg cloth	[74]
plain woven	none	fiber (0.1 mm to 1.5 mm)	20.6–59.9 mg MNPs/kg cloth	[74]
teabag	abrasive wear (particle collision)	fragments (N.A.)	0.9–1.1 mg MNPs/teabag/filter	[72]

Table 1. *Cont.*

Production	Primary Mechanical Abrasion (Force Analysis)	Secondary MNPs	Release Tendency	Reference
teabag	abrasive wear (particle collision)	fragments (N.A.)	0.7–1.0 MNPs/teabag/filter	[72]
water bottle	adhesive wear (compression)	fragments (N.A.)	112–553 MNPs/L/cycle	[59]
teabag	abrasive wear (particle collision)	fragments (N.A.)	94% of samples released MNPs	[75]
teabag	abrasive wear (particle collision)	fiber (N.A.)	53.4–80.1 MNPs/kg teabag	[30]
fabric production	adhesive wear (compression and twisting)	fiber (N.A.)	mean at 114 mg MNPs/kg fabric	[76]
rope	adhesive wear (compression and twisting)	fragments (N.A.)	11–1052 µg MNPs/m rope	[77]
bowls	N.A.	N.A.	331–898 MNPs/treatment	[78]
facial scrub	abrasive wear (shear stress)	fragments (nano-sized)	up to 10^{11} MNPs/4g samples	[79]

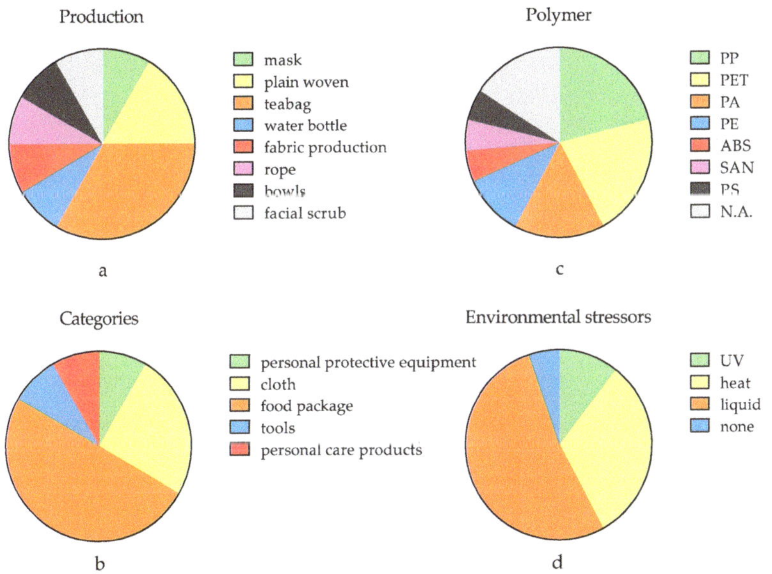

Figure 5. The distribution of the production types (**a**), categories according to the usage scenario (**b**), polymers (**c**) and environmental stressors (**d**) in the analyzed literature. The abbreviations: Polyamide (PA), Polyethylene (PE), Polyethylene Terephthalate (PET), acrylo-nitrile–butadiene–styrene (ABS), polypropylene (PP), melamine, polyethylene (PE), polystyrene (PS), and styrene–acrylonitrile (SAN); Not available in original references (N.A.).

There are limited data on the impacts of human exposure to MNPs, despite studies being conducted on the fragmentation of plastic items in daily life. According to estimates based solely on paper cups, the average chronic daily intake was 0.03 mg MNPs/kg body weight per day [80]. This value is much higher than the quantity of microplastics that a person would normally consume by eating seafood [81]. In other words, it is preferable to check your apron for tears when preparing seafood than to be concerned about consuming an excessive amount of MNPs from your meal. Due to variations in measurements, such as the number of particles or the mass of the particles, which result in variances in the calculations, it is currently challenging to create a broadly applicable baseline of MNPs contamination from the fragmentation process. Alarmingly, recent simulations have demonstrated that low levels of shear force (4.0 kJ/L) are also capable of generating large quantities of MNPs, and that these smaller plastics can trigger apoptosis in fish cells [79]. In order to establish a baseline for chronic human exposure, it is necessary to determine the rate at which typical plastic products generate MNPs during usage using a standardized experimental setup.

5.2. Long-Term Exposure

In comparison with chronic and direct exposure, the long-term exposures from indirect pathways are more subtle, but their health risks to humans are significant. Based on the limited evidence, we confirmed that waste recycling, landfill and municipal activities represented pathways for the long-term exposure of humans to MNPs (Table 2). Waste recycling and landfill are the most typical forms of anthropogenic fragmentation of plastics by mechanical forces, as well as typical sources of pollutant emissions. In the first place, they are hazardous workplaces with a high abundance of MNPs, and onsite workers and nearby inhabitants are exposed to huge quantities of these particles, which are suspended during the fragmentation process and are inhaled via respiration. In addition to direct contact, wastewater, sludge and even drinking water are primary pathways for the transportation of secondary MNPs. The presence of MNPs in tap water has been confirmed by numerous studies and can be traced back to contaminated water sources, insufficient wastewater treatment and atmospheric deposition, which are thought to be the primary causes of tap-water pollution. There are only limited studies on the MNPs that are created when tap-water networks (e.g., kettles) break down [64,82,83]. As pipelines age, the water conveyance process may become a more visible source of MNPs contamination compared with the water source.

The usage of polymer composite materials is anticipated to enhance the danger of human exposure to microplastics in some areas where mechanical abrasion is substantial (e.g., highways and stadiums) [73,84,85]. Even with low-energy abrasion, MNPs can be released gradually. It was estimated that abrasive wear of shoe outsoles can produce up to 0.9 MNPs/linear meter/runner in trail-running events [54]. Long-term mechanical fragmentation mostly releases MNPs via adhesive wear and slow-acting compressive stresses. Therefore, the overall release of MNPs is significant compared with acute exposure pathways since these effects typically occur under conditions of significant environmental stress (e.g., fully open outdoor environments and fluid-filled pipes). The management of long-term exposure pathways will become more challenging as human development and outdoor activity increase together with the usage of novel materials. For environmental managers, reducing the long-term exposure pathways means not only controlling the sources of pollution but also monitoring and controlling the operation of various production and living activities.

Table 2. Potential pathways for long-term exposure of humans to MNPs.

Production	Pathways	Pollution Matrix	Environmental Stress	Primary Mechanical Abrasion	Release Tendency	Reference
water bottle	waste recycling	waste water/sludge	natural conditions	shearing force	967–24,798 MNPs/L; 773–1450 MNPs/kg	[86]
household plastic	waste recycling	waste water	natural conditions with UV	shearing force	110–200,000 MNPs/L	[87]
household plastic	landfill	surface soil	natural conditions with UV	compression	56–122 MNPs/5 cm^2 samples	[88]
construction material	municipal activities	surface dust	liquid/heat	compression	0.53 ± 0.15 g MNPs/m^2	[73]
plastic kettles	municipal activities	drinking water	liquid	compression	N.A. [a]	[89]

[a] Not available in original references (N.A.).

6. Outlook

6.1. Toxicological Approach Based on the Full Life Cycle of Plastic Products

Understanding the ecological risks of plastics is hampered by the compartmentalization of daughter productions, e.g., MNPs in aging and virgin plastic products in current studies. In indoor-exposure experiments, either brand-new polymer particles are used as exposure models or particles of already highly aged products are used as a substitute for real-world conditions. The aging of plastics, however, is a dynamic process and does not progress at a constant rate. The characterization and toxicological impacts of secondary pollutants, including MNPs and POPs, emitted at various stages of plastic aging must, therefore, be studied. This will make it possible to properly contextualize toxicological research using plastics and will serve as a foundation for future screenings of potential hazards of environmental contamination due to plastic waste on our planet.

6.2. Integration of Aging Evaluation into Risk Assessment Frameworks

For a long period, the evaluation of the aging degree was only used to assess the performance and service life of plastic products. There is a lack of consideration for these processes in the current risk assessment framework for MNPs, despite the strong relationship between the rate of aging, its byproducts and the health hazards connected to them. The health concerns that plastics may present to species, including humans, will be grossly understated by this omission. Therefore, aging degree should be taken into account as a significant indicator of forecast risk in future risk assessment frameworks. Additionally, quantitative methods should be employed to evaluate the contribution of mechanical abrasion and wear to the progression of aging in addition to studies of environmental stressors. The risk of human exposure to plastic products at various phases of usage and service life should be evaluated on this basis. We urge the investigation of new, less instrument-dependent imaging techniques or surface physics that may be utilized to rapidly determine the level of aging and contaminant release.

6.3. Establishing Early Warning Signals of MNPs Released from Plastic Items

The release of MNPs from plastic products is significantly variable depending on the polymer composition and processing techniques used. The release speed and the morphological properties of secondary MNPs are also influenced by variations in usage scenarios and the forms of mechanical fragmentation. Therefore, aging prediction models for common plastic items under various usage patterns, intensities and predicted lifetimes should be created in the future, and physicochemical markers that accurately indicate the progression of aging should be screened. Accelerated aging studies carried out indoors should be used to quantify the release rate of MNPs per unit of time or number of uses. We suggest that the fragmentation potential and environmental aging performance of various plastic materials be included in plastic risk assessment criteria together with the ongoing enhancement of the human health risk evaluation benchmarks for MNPs. This will be beneficial in assessing the hazards of long-term plastic exposure and can be used to direct infrastructure building, material choice and protective strategies for infrastructures that frequently come into contact with people.

7. Conclusions

Our current work has revealed hidden pathways regarding human exposure to MNPs via a literature review. The mechanical fragmentation of plastic products during daily use is a critical contributor to the secondary MNPs generated from abrasive and adhesive wearing. The physical features of those MNPs are determined by the size categories and mechanisms of fragmentation, which are closely related to their toxic effects and environmental behaviors. Acute and long-term pathways can be distinguished, with studies typically ignoring the latter. We proposed that future studies should use a toxicological approach based on the full life cycle of plastic products and include an aging evaluation in frameworks for risk assessment.

Author Contributions: Conceptualization, Y.Y. and L.S.; methodology, N.C.; writing—original draft preparation, Y.Y.; writing—review and editing, L.S. and N.C.; visualization, Y.Y.; project administration, L.S. All authors have read and agreed to the published version of the manuscript.

Funding: This research was funded by the MNR Key Laboratory of Eco-Environment Science and Technology, China (MEEST202201), Opening Foundation of Guangxi Key Laboratory of Marine Environmental Change and Disaster in Beibu Gulf, Beibu Gulf University, Qinzhou, China (No. 2021KF05).

Acknowledgments: This work was supported by the MNR Key Laboratory of Eco-Environment Science and Technology, China and Opening Foundation of Guangxi Key Laboratory of Marine Environmental Change and Disaster in Beibu Gulf, Beibu Gulf University, Qinzhou, China.

Conflicts of Interest: The authors declare no conflict of interest.

References

1. Borrelle, S.B.; Ringma, J.; Law, K.L.; Monnahan, C.C.; Lebreton, L.; McGivern, A.; Murphy, E.; Jambeck, J.; Leonard, G.H.; Hilleary, M.A.; et al. Predicted growth in plastic waste exceeds efforts to mitigate plastic pollution. *Science* **2020**, *369*, 1515–1518. [CrossRef] [PubMed]
2. Bucci, K.; Tulio, M.; Rochman, C.M. What is known and unknown about the effects of plastic pollution: A meta-analysis and systematic review. *Ecol. Appl.* **2020**, *30*, 16. [CrossRef]
3. Geyer, R.; Jambeck, J.R.; Law, K.L. Production, use, and fate of all plastics ever made. *Sci. Adv.* **2017**, *3*, e1700782. [CrossRef] [PubMed]
4. Varg, J.E.; Svanbäck, R. Multi stress system: Microplastics in freshwater and their effects on host microbiota. *Sci. Total. Environ.* **2023**, *856*, 159106. [CrossRef] [PubMed]
5. Seltenrich, N. New link in the food chain? marine plastic pollution and seafood safety. *Environ. Health Perspect.* **2015**, *123*, A34–A41. [CrossRef] [PubMed]
6. Bergami, E.; Bocci, E.; Vannuccini, M.L.; Monopoli, M.; Salvati, A.; Dawson, K.A.; Corsi, I. Nano-sized polystyrene affects feeding, behavior and physiology of brine shrimp Artemia franciscana larvae. *Ecotox. Environ. Safe.* **2016**, *123*, 18–25. [CrossRef]
7. Wright, S.L.; Thompson, R.C.; Galloway, T.S. The physical impacts of microplastics on marine organisms: A review. *Environ. Pollut.* **2013**, *178*, 483–492. [CrossRef]
8. Li, B.W.; Liang, W.W.H.; Liu, Q.X.; Fu, S.J.; Ma, C.Z.; Chen, Q.Q.; Su, L.; Craig, N.J.; Shi, H.H. Fish ingest microplastics unintentionally. *Environ. Sci. Technol.* **2021**, *55*, 10471–10479. [CrossRef]
9. Kramm, J.; Völker, C.; Wagner, M. Superficial or substantial: Why care about microplastics in the anthropocene? *Environ. Sci. Technol.* **2018**, *52*, 3336–3337. [CrossRef]
10. Frias, J.P.G.L.; Nash, R. Microplastics: Finding a consensus on the definition. *Mar. Pollut. Bull.* **2019**, *138*, 145–147. [CrossRef]
11. Van Cauwenberghe, L.; Janssen, C.R. Microplastics in bivalves cultured for human consumption. *Environ. Pollut.* **2014**, *193*, 65–70. [CrossRef]
12. Collard, F.; Ask, A. Plastic ingestion by Arctic fauna: A review. *Sci. Total. Environ.* **2021**, *786*, 147462. [CrossRef]
13. Danopoulos, E.; Jenner, L.C.; Twiddy, M.; Rotchell, J.M. Microplastic contamination of seafood intended for human consumption: A systematic review and meta–analysis. *Environ. Health Perspect.* **2020**, *128*, 32. [CrossRef]
14. Su, L.; Deng, H.; Li, B.; Chen, Q.; Pettigrove, V.; Wu, C.; Shi, H. The occurrence of microplastic in specific organs in commercially caught fishes from coast and estuary area of east China. *J. Hazard. Mater.* **2019**, *365*, 716–724. [CrossRef]
15. Zhang, Q.; Xu, E.G.; Li, J.; Chen, Q.; Ma, L.; Zeng, E.Y.; Shi, H. A Review of Microplastics in Table Salt, Drinking Water, and Air: Direct Human Exposure. *Environ. Sci. Technol.* **2020**, *54*, 3740–3751. [CrossRef]
16. Wright, S.; Borm, P.J.A. Applying Existing Particle Paradigms to Inhaled Microplastic Particles. *Front. Public Health* **2022**, *10*, 868822. [CrossRef]
17. Roychand, R.; Pramanik, B.K. Identification of micro-plastics in Australian road dust. *J. Environ. Chem. Eng.* **2020**, *8*, 103647. [CrossRef]
18. Su, L.; Nan, B.; Craig, N.J.; Pettigrove, V. Temporal and spatial variations of microplastics in roadside dust from rural and urban Victoria, Australia: Implications for diffuse pollution. *Chemosphere* **2020**, *252*, 126567. [CrossRef]
19. Rani-Borges, B.; Queiroz, L.G.; Pradd, C.C.A.; de Melo, E.C.; de Moraes, B.R.; Ando, R.A.; de Paiva, T.C.B.; Pompeo, M. Exposure of the amphipod Hyalella azteca to microplastics. A study on subtoxic responses and particle biofragmentation. *Aquat. Toxicol.* **2023**, *258*, 106516. [CrossRef] [PubMed]
20. ter Halle, A.; Ladirat, L.; Martignac, M.; Mingotaud, A.F.; Boyron, O.; Perez, E. To what extent are microplastics from the open ocean weathered? *Environ. Pollut.* **2017**, *227*, 167–174. [CrossRef]
21. Zheng, Y.; Hamed, M.; De-la-Torre, G.E.; Frias, J.; Jong, M.-C.; Kolandhasamy, P.; Chavanich, S.; Su, L.; Deng, H.; Zhao, W.; et al. Holes on surfaces of the weathered plastic fragments from coastal beaches. *Mar. Pollut. Bull.* **2023**, *193*, 115180. [CrossRef] [PubMed]

22. Acosta-Coley, I.; Olivero-Verbel, J. Microplastic resin pellets on an urban tropical beach in Colombia. *Environ. Monit. Assess.* **2015**, *187*, 435. [CrossRef] [PubMed]
23. Hays, H.; Cormons, G. Plastic particles found in tern pellets, on coastal beaches and at factory sites. *Mar. Pollut. Bull.* **1974**, *5*, 44–46. [CrossRef]
24. Bour, A.; Sturve, J.; Hojesjo, J.; Almroth, B.C. Microplastic vector effects: Are fish at risk when exposed via the trophic chain? *Front. Environ. Sci.* **2020**, *8*, 10. [CrossRef]
25. Guan, J.; Qi, K.; Wang, J.; Wang, W.; Wang, Z.; Lu, N.; Qu, J. Microplastics as an emerging anthropogenic vector of trace metals in freshwater: Significance of biofilms and comparison with natural substrates. *Water Res.* **2020**, *184*, 116205. [CrossRef]
26. ter Halle, A.; Ladirat, L.; Gendre, X.; Goudouneche, D.; Pusineri, C.; Routaboul, C.; Tenailleau, C.; Duployer, B.; Perez, E. Understanding the fragmentation pattern of marine plastic debris. *Environ. Sci. Technol.* **2016**, *50*, 5668–5675. [CrossRef] [PubMed]
27. Garvey, C.J.; Impéror-Clerc, M.; Rouzière, S.; Gouadec, G.; Boyron, O.; Rowenczyk, L.; Mingotaud, A.F.; ter Halle, A. Molecular-scale understanding of the embrittlement in polyethylene ocean debris. *Environ. Sci. Technol.* **2020**, *54*, 11173–11181. [CrossRef]
28. Shen, M.; Zeng, Z.; Song, B.; Yi, H.; Hu, T.; Zhang, Y.; Zeng, G.; Xiao, R. Neglected microplastics pollution in global COVID-19: Disposable surgical masks. *Sci. Total. Environ.* **2021**, *790*, 148130. [CrossRef]
29. Hussain, K.A.; Romanova, S.; Okur, I.; Zhang, D.; Kuebler, J.; Huang, X.; Wang, B.; Fernandez-Ballester, L.; Lu, Y.; Schubert, M.; et al. Assessing the Release of Microplastics and Nanoplastics from Plastic Containers and Reusable Food Pouches: Implications for Human Health. *Environ. Sci. Technol.* **2023**, *57*, 9782–9792. [CrossRef]
30. Afrin, S.; Rahman, M.M.; Akbor, M.A.; Siddique, M.A.B.; Uddin, M.K.; Malafaia, G. Is there tea complemented with the appealing flavor of microplastics? A pioneering study on plastic pollution in commercially available tea bags in Bangladesh. *Sci. Total. Environ.* **2022**, *837*, 155833. [CrossRef]
31. Arp, H.P.H.; Kühnel, D.; Rummel, C.; MacLeod, M.; Potthoff, A.; Reichelt, S.; Rojo-Nieto, E.; Schmitt-Jansen, M.; Sonnenberg, J.; Toorman, E.; et al. Weathering plastics as a planetary boundary threat: Exposure, fate, and hazards. *Environ. Sci. Technol.* **2021**, *55*, 7246–7255. [CrossRef]
32. Bourne, W.R.P. Nylon netting as a hazard to birds. *Mar. Pollut. Bull.* **1977**, *8*, 75–76. [CrossRef]
33. Satoto, R.; Subowo, W.S.; Yusiasih, R.; Takane, Y.; Watanabe, Y.; Hatakeyama, T. Weathering of high-density polyethylene in different latitudes. *Polym. Degrad. Stabil.* **1997**, *56*, 275–279. [CrossRef]
34. Born, M.P.; Brüll, C. From model to nature—A review on the transferability of marine (micro-) plastic fragmentation studies. *Sci. Total. Environ.* **2022**, *811*, 151389. [CrossRef] [PubMed]
35. Khosrovyan, A.; Kahru, A. Evaluation of the potential toxicity of UV-weathered virgin polyamide microplastics to non-biting midge Chironomus riparius. *Environ. Pollut.* **2021**, *287*, 117334. [CrossRef]
36. Klein, K.; Hof, D.; Dombrowski, A.; Schweyen, P.; Dierkes, G.; Ternes, T.; Schulte-Oehlmann, U.; Oehlmann, J. Enhanced in vitro toxicity of plastic leachates after UV irradiation. *Water Res.* **2021**, *199*, 117203. [CrossRef]
37. Alimi, O.S.; Claveau-Mallet, D.; Kurusu, R.S.; Lapointe, M.; Bayen, S.; Tufenkji, N. Weathering pathways and protocols for environmentally relevant microplastics and nanoplastics: What are we missing? *J. Hazard. Mater.* **2022**, *423*, 126955. [CrossRef]
38. Hayes, M.D.; Edwards, D.B.; Shah, A.R. 1. Introduction. In *Fractography in Failure Analysis of Polymers*; Hayes, M.D., Edwards, D.B., Shah, A.R., Eds.; William Andrew Publishing: Oxford, UK, 2015; pp. 1–5.
39. Gewert, B.; Plassmann, M.M.; MacLeod, M. Pathways for degradation of plastic polymers floating in the marine environment. *Environ. Sci. Process. Impacts* **2015**, *17*, 1513–1521. [CrossRef]
40. Fallahi, H.; Taheri-Behrooz, F.; Asadi, A. Nonlinear Mechanical Response of Polymer Matrix Composites: A Review. *Polym. Rev.* **2020**, *60*, 42–85. [CrossRef]
41. Andrade, J.; Fernández-González, V.; López-Mahía, P.; Muniategui, S. A low-cost system to simulate environmental microplastic weathering. *Mar. Pollut. Bull.* **2019**, *149*, 110663. [CrossRef]
42. Menzel, T.; Meides, N.; Mauel, A.; Mansfeld, U.; Kretschmer, W.; Kuhn, M.; Herzig, E.M.; Altstädt, V.; Strohriegl, P.; Senker, J.; et al. Degradation of low-density polyethylene to nanoplastic particles by accelerated weathering. *Sci. Total. Environ.* **2022**, *826*, 154035. [CrossRef] [PubMed]
43. Julienne, F.; Lagarde, F.; Bardeau, J.-F.; Delorme, N. Thin polyethylene (LDPE) films with controlled crystalline morphology for studying plastic weathering and microplastic generation. *Polym. Degrad. Stabil.* **2022**, *195*, 109791. [CrossRef]
44. Julienne, F.; Lagarde, F.; Delorme, N. Influence of the crystalline structure on the fragmentation of weathered polyolefines. *Polym. Degrad. Stabil.* **2019**, *170*, 8. [CrossRef]
45. Min, K.; Cuiffi, J.D.; Mathers, R.T. Ranking environmental degradation trends of plastic marine debris based on physical properties and molecular structure. *Nat. Commun.* **2020**, *11*, 727. [CrossRef]
46. Liu, S.; Huang, W.; Yang, J.; Xiong, Y.; Huang, Z.; Wang, J.; Cai, T.; Dang, Z.; Yang, C. Formation of environmentally persistent free radicals on microplastics under UV irradiations. *J. Hazard. Mater.* **2023**, *453*, 131277. [CrossRef] [PubMed]
47. Barnes, D.K.; Galgani, F.; Thompson, R.C.; Barlaz, M. Accumulation and fragmentation of plastic debris in global environments. *Philos. Trans. R. Soc. B* **2009**, *364*, 1985–1998. [CrossRef]
48. Andrady, A.L. Weathering and fragmentation of plastic debris in the ocean environment. *Mar. Pollut. Bull.* **2022**, *180*, 113761. [CrossRef]
49. Song, Y.K.; Hong, S.H.; Eo, S.; Shim, W.J. The fragmentation of nano- and microplastic particles from thermoplastics accelerated by simulated-sunlight-mediated photooxidation. *Environ. Pollut.* **2022**, *311*, 119847. [CrossRef]

50. Hueper, W.C. Cancer Induction by polyurethan and polysilicone plastics. *J. Natl. Cancer Inst.* **1964**, *33*, 1005–1027.
51. Thomson, J.G.; Kaschula, R.O.; Macdonald, R.R. Asbestos as a modern urban hazard. *S. Afr. Med. J.* **1963**, *37*, 77–81.
52. Takahashi, Y.; Tanaka, K.; Kajiwara, T.; Suzuki, G.; Osako, M.; Kuramochi, H. Cross-sectional microstructural analysis to evaluate the crack growth pattern of weathered marine plastics. *Chemosphere* **2023**, *331*, 138794. [CrossRef] [PubMed]
53. Deng, H.; Su, L.; Zheng, Y.F.; Du, F.N.; Liu, Q.X.; Zheng, J.; Zhou, Z.W.; Shi, H.H. Crack Patterns of Environmental Plastic Fragments. *Environ. Sci. Technol.* **2022**, *56*, 6399–6414. [CrossRef] [PubMed]
54. Forster, N.A.; Wilson, S.C.; Tighe, M.K. Trail running events contribute microplastic pollution to conservation and wilderness areas. *J. Environ. Manag.* **2023**, *331*, 117304. [CrossRef] [PubMed]
55. Eisentraut, P.; Dumichen, E.; Ruhl, A.S.; Jekel, M.; Albrecht, M.; Gehde, M.; Braun, U. Two birds with one stone—Fast and simultaneous analysis of microplastics: Microparticles derived from thermoplastics and tire wear. *Environ. Sci. Technol.* **2018**, *5*, 608–613. [CrossRef]
56. Leads, R.R.; Weinstein, J.E. Occurrence of tire wear particles and other microplastics within the tributaries of the Charleston Harbor Estuary, South Carolina, USA. *Mar. Pollut. Bull.* **2019**, *145*, 569–582. [CrossRef]
57. Parker, B.W.; Beckingham, B.A.; Ingram, B.C.; Ballenger, J.C.; Weinstein, J.E.; Sancho, G. Microplastic and tire wear particle occurrence in fishes from an urban estuary: Influence of feeding characteristics on exposure risk. *Mar. Pollut. Bull.* **2020**, *160*, 111539. [CrossRef]
58. Winkler, A.; Santo, N.; Ortenzi, M.A.; Bolzoni, E.; Bacchetta, R.; Tremolada, P. Does mechanical stress cause microplastic release from plastic water bottles? *Water Res.* **2019**, *166*, 115082. [CrossRef]
59. Tvisha, S. Generation of microplastics from the opening and closing of disposable plastic water bottles. *J. Water Health* **2021**, *19*, 488–498. [CrossRef]
60. Su, L.; Du, F.; Sun, C.; Shi, H. Linking the physical and chemical characteristics of single small microplastics or nanoplastics via photolithographic silicon substrates. *Anal. Methods* **2022**, *14*, 1547–1552. [CrossRef]
61. Deen, N.G.; Kriebitzsch, S.H.L.; van der Hoef, M.A.; Kuipers, J.A.M. Direct numerical simulation of flow and heat transfer in dense fluid-particle systems. *Chem. Eng. Sci.* **2012**, *81*, 329–344. [CrossRef]
62. Forward, K.M.; Lacks, D.J.; Sankaran, R.M. Charge segregation depends on particle size in triboelectrically charged granular materials. *Phys. Rev. Lett.* **2009**, *102*, 028001. [CrossRef] [PubMed]
63. Kwak, J.I.; An, Y.-J. Microplastic digestion generates fragmented nanoplastics in soils and damages earthworm spermatogenesis and coelomocyte viability. *J. Hazard. Mater.* **2021**, *402*, 124034. [CrossRef] [PubMed]
64. Luo, Y.; Al Amin, M.; Gibson, C.T.; Chuah, C.; Tang, Y.; Naidu, R.; Fang, C. Raman imaging of microplastics and nanoplastics generated by cutting PVC pipe. *Environ. Pollut.* **2022**, *298*, 118857. [CrossRef] [PubMed]
65. Alekseev, I.; Miroslavov, A. Post-effects of radioactive decay in magnetite nano-crystals labelled with Auger- and internal conversion electron-emitters, alpha- and beta decay radionuclides. *Radiat. Phys. Chem.* **2020**, *177*, 109160. [CrossRef]
66. Wang, W.; Zhang, J.; Qiu, Z.; Cui, Z.; Li, N.; Li, X.; Wang, Y.; Zhang, H.; Zhao, C. Effects of polyethylene microplastics on cell membranes: A combined study of experiments and molecular dynamics simulations. *J. Hazard. Mater.* **2022**, *429*, 128323. [CrossRef]
67. Das, A. The emerging role of microplastics in systemic toxicity: Involvement of reactive oxygen species (ROS). *Sci. Total. Environ.* **2023**, *895*, 165076. [CrossRef]
68. Koelmans, A.A.; Bakir, A.; Burton, G.A.; Janssen, C.R. Microplastic as a vector for chemicals in the aquatic environment: Critical review and model-supported reinterpretation of empirical studies. *Environ. Sci. Technol.* **2016**, *50*, 3315–3326. [CrossRef]
69. Wang, Z.; Dong, H.; Wang, Y.; Ren, R.; Qin, X.; Wang, S. Effects of microplastics and their adsorption of cadmium as vectors on the cladoceran Moina monogolica Daday: Implications for plastic-ingesting organisms. *J. Hazard. Mater.* **2020**, *400*, 123239. [CrossRef]
70. Andrady, A.L. The plastic in microplastics: A review. *Mar. Pollut. Bull.* **2017**, *119*, 12–22. [CrossRef]
71. Zhang, X.; Li, Y.; Ouyang, D.; Lei, J.; Tan, Q.; Xie, L.; Li, Z.; Liu, T.; Xiao, Y.; Farooq, T.H. Systematical review of interactions between microplastics and microorganisms in the soil environment. *J. Hazard. Mater.* **2021**, *418*, 126288. [CrossRef]
72. Cella, C.; La Spina, R.; Mehn, D.; Fumagalli, F.; Ceccone, G.; Valsesia, A.; Gilliland, D. Detecting Micro- and Nanoplastics Released from Food Packaging: Challenges and Analytical Strategies. *Polymers* **2022**, *14*, 1238. [CrossRef] [PubMed]
73. Enfrin, M.; Myszka, R.; Giustozzi, F. Paving roads with recycled plastics: Microplastic pollution or eco-friendly solution? *J. Hazard. Mater.* **2022**, *437*, 129334. [CrossRef] [PubMed]
74. Choi, S.; Kwon, M.; Park, M.-J.; Kim, J. Analysis of Microplastics Released from Plain Woven Classified by Yarn Types during Washing and Drying. *Polymers* **2021**, *13*, 2988. [CrossRef] [PubMed]
75. Tingna, M.; Jiahua, W.; Xiaofeng, X.; Jingwen, L.; Qiaocong, L.; Huang, D.; Xiaodan, L.; Fuwei, P. Identification and Evaluation of Microplastics from Tea Filter Bags Based on Raman Imaging. *Foods* **2022**, *11*, 2871. [CrossRef]
76. Lant, N.J.; Hayward, A.S.; Peththawadu, M.M.D.; Sheridan, K.J.; Dean, J.R. Microfiber release from real soiled consumer laundry and the impact of fabric care products and washing conditions. *PLoS ONE* **2020**, *15*, e0233332. [CrossRef]
77. Napper, I.E.; Wright, L.S.; Barrett, A.C.; Parker-Jurd, F.N.F.; Thompson, R.C. Potential microplastic release from the maritime industry: Abrasion of rope. *Sci. Total. Environ.* **2022**, *804*, 150155. [CrossRef]
78. Jander, J.; Hummel, D.; Stuermer, S.; Monteleone, A.; Neumaier, T.; Broghammer, F.; Lewin-Kretzschmar, U.; Brock, T.; Knoll, M.; Fath, A.S. Release of Microplastics from Reusable Kitchen Plasticware and Generation of Thermal Potential Toxic Degradation Products in the Oven. *Appl. Sci.* **2022**, *12*, 2535. [CrossRef]

79. Enfrin, M.; Lee, J.; Gibert, Y.; Basheer, F.; Kong, L.; Dumée, L.F. Release of hazardous nanoplastic contaminants due to microplastics fragmentation under shear stress forces. *J. Hazard. Mater.* **2020**, *384*, 121393. [CrossRef]
80. Joseph, A.; Parveen, N.; Ranjan, V.P.; Goel, S. Drinking hot beverages from paper cups: Lifetime intake of microplastics. *Chemosphere* **2023**, *317*, 137844. [CrossRef]
81. Li, J.; Green, C.; Reynolds, A.; Shi, H.; Rotchell, J.M. Microplastics in mussels sampled from coastal waters and supermarkets in the United Kingdom. *Environ. Pollut.* **2018**, *241*, 35–44. [CrossRef]
82. Monira, S.; Roychand, R.; Bhuiyan, M.A.; Pramanik, B.K. Role of water shear force for microplastics fragmentation into nanoplastics. *Environ. Res.* **2023**, *237*, 116916. [CrossRef] [PubMed]
83. Monira, S.; Roychand, R.; Hai, F.I.; Bhuiyan, M.; Dhar, B.R.; Pramanik, B.K. Nano and microplastics occurrence in wastewater treatment plants: A comprehensive understanding of microplastics fragmentation and their removal. *Chemosphere* **2023**, *334*, 139011. [CrossRef] [PubMed]
84. de Haan, W.P.; Quintana, R.; Vilas, C.; Cózar, A.; Canals, M.; Uviedo, O.; Sanchez-Vidal, A. The dark side of artificial greening: Plastic turfs as widespread pollutants of aquatic environments. *Environ. Pollut.* **2023**, *334*, 122094. [CrossRef] [PubMed]
85. Yin, M.; Yang, L.; Tang, Y.; Liang, Y.; Wang, Q.; Su, L.; Liu, J.; Zhuang, Y.; Li, D. Microplastics released from artificial turf applied as hedge walls: Their aging-induced properties and uptake by grass carp, mussels and earthworms. *Process. Saf. Environ.* **2023**, *174*, 53–62. [CrossRef]
86. Guo, Y.; Xia, X.; Ruan, J.; Wang, Y.; Zhang, J.; LeBlanc, G.A.; An, L. Ignored microplastic sources from plastic bottle recycling. *Sci. Total. Environ.* **2022**, *838*, 156038. [CrossRef]
87. Suzuki, G.; Uchida, N.; Le Huu, T.; Tanaka, K.; Matsukami, H.; Kunisue, T.; Takahashi, S.; Pham Hung, V.; Kuramochi, H.; Osako, M. Mechanical recycling of plastic waste as a point source of microplastic pollution. *Environ. Pollut.* **2022**, *303*, 119114. [CrossRef]
88. Sholokhova, A.; Denafas, G.; Ceponkus, J.; Kriukiene, R. Microplastics Release from Conventional Plastics during Real Open Windrow Composting. *Sustainability* **2023**, *15*, 758. [CrossRef]
89. Shi, Y.; Li, D.; Xiao, L.; Sheerin, E.D.; Mullarkey, D.; Yang, L.; Bai, X.; Shvets, I.V.; Boland, J.J.; Wang, J.J. The influence of drinking water constituents on the level of microplastic release from plastic kettles. *J. Hazard. Mater.* **2022**, *425*, 127997. [CrossRef]

MDPI AG
Grosspeteranlage 5
4052 Basel
Switzerland
Tel.: +41 61 683 77 34

Toxics Editorial Office
E-mail: toxics@mdpi.com
www.mdpi.com/journal/toxics

www.ingramcontent.com/pod-product-compliance
Lightning Source LLC
Chambersburg PA
CBHW062052220926
51850CB00010B/2747
* 9 7 8 3 7 2 5 8 6 2 4 2 9 *